SPACE, TIME, CONSCIOUSNESS, OUT-OF-BODY EXPERIENCES AND CREATING THE FUTURE

SPACE, TIME, CONSCIOUSNESS, OUT-OF-BODY EXPERIENCES AND CREATING THE FUTURE

Treilor Banks

Copyright © 2022 by Wayne Van

All rights reserved.

Library of Congress Control Number: 2022914662

ISBN: 979-8-218-06791-5

DEDICATION

I dedicate this book to my wife Linda, my rock.

I also dedicate this book to all who wonder and seek understanding.

TABLE OF CONTENTS

Introduction — 9

Chapter 1
Breaking Through — 11

Chapter 2
Turning Off Our Sensors to Imagine the Future — 14

Chapter 3
Imagining the Future — 18

Chapter 4
Anatomy of an Event — 21

Chapter 5
Inside Events and the Creation of Out-Of-Body Experiences — 29

Chapter 6
Creating Waves with a Circle — 34

Chapter 7
Euler's Number: The Constant "e" — 53

Chapter 8
The Imaginary Circle and Euler's Formula — 56

Chapter 9
Hyperbolic Spacetime — 61

Chapter 10
Creating Waves in Hyperbolic Time — 72

Chapter 11
Having an OBE — 80

Chapter 12
Outside Events and Creating the Future — 83

Chapter 13
Requirements for Creating Outside Events   96

Chapter 14
Evidence for Outside Events   98

Chapter 15
Final Thoughts   100

Author Biography   101

INTRODUCTION

It has been quite a journey. From beginning to have spontaneous out-of-body experiences, or OBEs, over thirty-five years ago, to finding a method to induce OBEs, to developing theories of not only OBEs but also how we can create the future. This book relates my story and my theories to the reader.

We exist in a four-dimensional spacetime universe: three spatial dimensions and one time dimension. The spatial dimensions (length, width, height) are orthogonal, or at right angles to one another, as we all know from our everyday experience. Our bodies are a collection of sensors (eyes, ears, skin, taste buds, nose) that connect us to the real world of the three-dimensional present through the brainwaves that they pass along to our consciousnesses.

The fourth dimension, time, is not at right angles to the three spatial dimensions we move around in. Time is not orthogonal to space. When we combine non-orthogonal time with orthogonal space, we end up with four-dimensional spacetime which is non-orthogonal. The non-orthogonal shape of spacetime is "hyperbolic" which simply means "related to hyperbolas". The shape of a nuclear reactor cooling tower is often a hyperbola. Albert Einstein's Theories (Special and General) of Relativity are related to the hyperbolic shape of spacetime. The future represents imaginary time because we have not yet experienced it in the present. But, as we shall see, our consciousnesses can move to and through the future and even change it. "Imaginary time" and "the future" are synonymous and will be used interchangeably in this book.

When we are in the present receiving information from our sensors, we can move through space. We do it every day. But, although we can experience the *flow* of time while we are in the present, we cannot visualize or move through time like we can move through space while we are in the present. To move through imaginary time, or the future, we have to turn off our sensors. In the present, we experience what is called "clock time". When we are able to turn off our sensors and move to the future, we experience imaginary time.

We can turn off our sensors by focusing or visualizing going down an imaginary cone, or funnel. If we can travel to and burst through the closed end of the funnel, we will find that our sensors have been turned off and that we can visualize and move through the future, where OBEs occur. It is an either/or scenario. If our sensors are active, or "on", we can experience events in the present but we can't visualize or move through the future. On the other hand, if our sensors are inactive, or "off", we can visualize and move through the future but we can't experience the present. We can either move through the future or we can move through the present.

OBEs occur in imaginary time, the future, when our sensors are off. To experience an OBE, we therefore have to find a way to turn off our sensors. I do this as noted above: by imagining going down a funnel until I reach and burst through the closed end. At that point, my sensors "let go" of me. I essentially "tunnel out" of my sensors. Since we think of our sensors as our bodies, you can say that a person can have an OBE when he or she tunnels out of their body to the future. When we are able to turn off our sensors for an extended period of time, we move to the future where we can create "inside event" brainwaves, or OBEs. We not only create OBEs in the future, we experience them there as well. Our sensors remain off during the entire OBE. An OBE ends when one or more of our sensors "turns on" by receiving external sensor-driven stimuli (e.g. - hearing a loud noise somewhere nearby) and we are called back to the present. OBEs are called "inside events" because they occur entirely within us. We have to turn off our sensors to *begin* an OBE and one or more of our sensors turning back on can *end* an OBE, but there are no sensors involved *while we are having* an OBE.

There are also "outside events". Outside events allow us to create the future. Outside events are not OBEs. We begin an outside event the same way we begin an inside event OBE: by shutting off our sensors. But then we move to the future and "deposit" brainwaves for a future event we want to create or participate in on what I call the "time escalator". We deposit the brainwaves of the event we want to experience on the time escalator by visualizing or imagining the event while we are in the future. Time only moves in one direction: from the future to the present. After we deposit the desired event brainwaves on the time escalator, they "trickle down" to the present where our sensors pick them up as they "roll off" the time escalator and we experience the event.

All participants in a specific future event are able to deposit their brainwaves on the time escalator by visualizing their desired outcome for the event after they have shut off their sensors and moved to the future. The brainwaves of the various participants who have deposited brainwaves describing their desired outcome for the event then combine either destructively (cancel each other out) or constructively (reinforce each other). The event that finally trickles off the time escalator into the present will represent the combination of all the participants' brainwaves. Unlike inside events, outside events involve our sensors which are needed to perceive the events which come to us when they arrive in the present from the future on the time escalator.

This book goes into detail on all of the above, and more.

Finally, this book involves some mathematics. The math is not hard and ranges from Arithmetic to High School Algebra I with some basic Trigonometry thrown in. All of the mathematics is explained from the ground up: readers who have not used their math skills for some time should have no trouble understanding the material. That being said, the chapters of the book have been arranged so that the mathematical material can be skipped without loss of the underlying concepts. Those readers who want to follow this route can skip Chapters 6 - 11 which develop a mathematical "explanation" for OBEs.

CHAPTER 1

BREAKING THROUGH

During the 1980s, I worked for a mid-America technology company. I regularly saw the company's financial statements in the course of my duties. The company was not doing well. As the company ran out of funds, it began laying off employees. It faced an impending bankruptcy in the very near future.

I had a stay-at-home wife and a young son with another child on the way. The doctor told us that the delivery was going to be by a relatively costly cesarean section. We were a young family and we did not have much money saved. Unfortunately, my company had financed its own health care program which was going to go away with the company.

Knowing my job was about to end and that we were looking at possible financial hardship, especially with the impending medical costs relating to the birth of our second child, created a lot of stress. I had trouble sleeping. I would wake up in the middle of the night wondering how my family and I were going to manage.

Then something strange began to happen. I would be asleep and all of a sudden, my head would begin to literally vibrate. It was as if someone was holding one of those pain relief muscle vibrators you can buy at the local drug store up against my head. After the vibrations began, I would hit myself against the side of my head and they would stop.

The vibrations would occur two or three times a week. Naturally, I was worried that something was very wrong with me. Did I have a brain tumor? What would happen if I didn't cause the vibrations to cease by hitting myself on the side of the head after they began?

One day I had a particularly bad day at work. My company had laid off a lot of people and there were very few of us corporate people left. It became clear to me that I was among the next to get the axe. My wife was facing an imminent delivery of our second child and I was not sure we were going to be able to pay for it. That night, the vibrations started. Instead of hitting myself on the side of my head to make them stop, I remember thinking, "Okay. Let's see what's on the other side!"

The vibrations continued unabated and I wasn't going to stop them. Then, all of a sudden, the vibrations stopped. Incredibly, I was suddenly at a college basketball game hundreds of miles away between two teams I had never heard of: Lafayette College and Lehigh College. I could see the names of the schools in the auditorium. I could see the fans yelling and screaming. The game was a barnburner. It went back and forth until the very end and was decided by a mere two points. It wasn't a dream. I was there! The reader may be thinking: "Well you had a vivid dream. Lots of people have vivid dreams that aren't real." Just wait.

The next morning, I went out to the driveway to pick up our paper. I took it inside and found the sports section and began to go through it, part of my weekend routine. I turned a page or two and then there it was. The headline read: "Lafayette and Lehigh down to the wire!" The article described the very game I had seen after the vibrations stopped. As mentioned, I had never even heard of Lafayette or Lehigh. I had never seen an article about either Pennsylvania college in our Kansas paper. No. It wasn't a dream.

I was flabbergasted. How could this have happened? I am a searcher. I have always tried to find out how things work and the reasons for why things happen. I couldn't let this go. I had to find an explanation. I went to a large bookstore that I frequented hoping to find something relating to my experience. (There was no internet in those days.) I ended up in a section named "Spiritual". There it was: a book on out-of-body experiences. After looking at the book I had my answer: I had had an out-of-body experience! Somehow, my consciousness had left my body and travelled from my bedroom in Kansas to the game between those two schools in Pennsylvania!

It is estimated that one in ten people have had an out-of-body experience (OBE) or something like it. I have a friend who ran track in college. We were talking one day and he told me that he was running laps one day and all of a sudden, he found himself outside of his body. He was above and looking down at his physical body. I don't know why he decided to tell me this. We weren't talking about OBEs. He asked me if I believed in out-of-body experiences. I said, "Oh yes." and then I told him about my experiences.

After the "basketball" episode, I had several more experiences in which my consciousness would leave my body after my head had started vibrating. Instead of being far away, though, I would normally initially be in my bedroom where I could stare down at my body lying in bed. I would then simply tell myself I wanted to leave the bedroom in a particular direction and I would move through a wall or closed door in that direction. When I say "move through" I do mean "through". I could will myself to go through a solid wall or closed door and it would happen. So much for solid walls and closed doors!

Literally anything I willed during an OBE could happen. I could leave my house and go outside and look around. I could clap my hands together and hear the noise they made without waking my wife. I remember one night I left my body and started staring at my wife while she was sleeping. She had bluish-green waves of what I would call "energy" flowing out of her head and disappearing into the distance.

After being outside of my body for a while, my consciousness would always return to my body. I did not will this to happen. It happened spontaneously. It's like I had a pass to be outside of my body, but only for so long.

Many times, after my consciousness had returned to my body, I would have a dream about an event which would come true the next day. Unfortunately, the dreams would never be about the lottery numbers I should play or a stock I should buy. The dreams would be something simple. For example, I might dream that I went to a grocery store and there was a clown inside handing out balloons to children. The next day I would be in a grocery store and there he was, a clown handing out balloons!

My theory is that I actually witness a future event during an OBE and then "see" that event during a dream after I return from the OBE. Seeing the future during/after an OBE is validation to me that the OBE was real. What are the chances that someone would see a clown handing out balloons during a dream then witness that same thing the next day? Pretty high unless the person had gone to the future and seen the event during an OBE. (This of course assumes that the person wasn't frequenting a grocery store where a clown regularly handed out balloons.)

I could not control the vibrations that led to the OBEs. I could not make them happen. But once they started, I knew if I let them continue they would result in my having an OBE.

My life improved. I was laid off from my job as I expected, but I was given a severance package that allowed us to pay for my wife's cesarean section and hospital stay during the birth of our second child. And luckily, I found another job within a couple of weeks from being laid off. And then the vibrations stopped. I believe the vibrations started due to the high amount of stress I was feeling in my life as discussed above. When the stress went away, the vibrations stopped. And when the vibrations stopped, so did the OBEs.

I am an analytical person. I can't stand not knowing how or why something works. I went to M.I.T. If you put a group of M.I.T. students in a room and give them a problem to solve, the first thing they will do is try to build a model, a mathematical solution to the problem that involves an equation or equations, that "explains" the solution to the problem.

So I set about trying to come up with a mathematical equation that described the OBEs I had experienced. I liked having OBEs and I did not want them to stop. I thought that if I could find a mathematical explanation for OBEs, I could use that explanation to show me how to continue having them. After much thought, I came up with a theory for the things I had experienced during an OBE: the ability for my consciousness to leave my body, to experience any event, to travel to the future, and so on.

CHAPTER 2

TURNING OFF OUR SENSORS TO IMAGINE THE FUTURE

You can never find a hill so high that, standing on the top of it, you will be able to view the world with yourself in it. Theories are not proof. Two of the world's greatest scientific achievements, Relativity Theory and Quantum Theory, are just that: theories. Relativity and Quantum theory were guesses which were at least partially confirmed by subsequent experiment. They have not been proven, per se, but they mathematically explain at least some events that take place in our world. Likewise, I am not attempting to *prove* OBEs. I am proposing a theory that will *explain* OBEs and the events I have encountered during them.

I know a little bit about quantum theory. It is very strange. An experiment in quantum theory traps an electron in a container from which it should not be able to escape. Nevertheless, the electron can escape the container and be found to exist outside of it. This is called "quantum tunneling." I thought, "Maybe I can tunnel out of my body just like the electron tunnels out of the box." I believe that the body is a container that imprisons us within its confines.

I soon discovered that visualizing that I was going down a funnel can actually lead to an OBE. A funnel is a tunnel in the shape of a cone which has one end closed. (We will use the words "funnel" and "tunnel" interchangeably in this book. The funnel's derivation will be mathematically addressed in Chapter 10.) What do I mean by "visualizing". When we visualize an event, we divorce ourselves from the physical world around us and use our mind's eye to see the event. There are really two worlds we exist in: the physical world and the world of our imagination. Using our imagination to visualize an event literally turns off our awareness of the physical world. The longer we are able to maintain a visualization, the more our awareness of the physical world declines.

Henry David Thoreau was an American naturalist, essayist, poet, and philosopher who lived between 1817 and 1862. In his book, "Walden", he famously stated during his sojourn at Walden Pond that, "My instinct tells me that my head is an organ for burrowing...". This may be even more true than he realized.

Our bodies are a collection of sensors: eyes (sight), ears (hearing), skin (touch), nose (smell), tongue (taste). These sensors are what present the physical world to our consciousness. Some people believe that our sensors (bodies) actually create our consciousnesses. I beg to differ. Implying that our sensors create "us", or our consciousnesses, implies that when our sensors stop working, "we", or our consciousnesses, cease to exist. I take the opposite view. Our sensors are a cage which contains our consciousness. An OBE temporarily removes the cage and allows us to conquer the constraints imposed by our sensors.

Our sensors, in conjunction with our brains, create what we know as "space". Every sensor we have helps determine and navigate our location in space. Sight allows us to see our spatial surroundings. Hearing allows us to orient ourselves in our spatial surroundings. Touch allows us to feel our spatial surroundings. Smell and taste allow us to differentiate our spatial surroundings. Removing all of our sensors would leave us completely unaware of the space which surrounds us. Part of our brain is called the "Parietal Lobe." The parietal lobe receives input from our sensors and creates our vision of the space which surrounds us. We will discuss this further later on in this book.

Scientists tell us that ninety-five percent of the mass and energy that make up our universe is "missing". They can't find it. I submit that this is the case because our bodies are not equipped with the sensors required to experience this missing mass and energy. It is not a problem of missing mass and energy, it is a problem of missing sensors.

Other than our sensors, which determine our concept of space, we have our imagination to visualize events which we want to happen in the future. When we visualize, we use our imaginations to create something, an event, in our mind's eye that is not discernable by our sensors. The things we create in our imaginations don't exist in the physical world, at least not when we first visualize them. You can create an object in the physical world and then show it to another person who will be able to observe it with his or her sensors. Not so with our imaginations. You can visualize an event with your imagination. You can describe your visualization to another person, but you cannot physically show it to him or her.

When we visualize an event, it involves something that occurs in future time. We can't visualize the present. Visualizing an event turns off our awareness of the physical world. In other words, visualizing turns off our sensors. Trying to visualize the present therefore "turns off" our awareness of the very present we are trying to visualize. Through our sensors, we experience present events that take place in space. Through our imaginations, we visualize future events that take place in time.

Our physical sensors give us information about the present. On the other hand, any time we visualize an event that has not yet taken place, we are attaching it to a future time in which it may occur. For example, we may visualize going to a party during the coming weekend.

Visualizing the future and sensing the present trade off with one another. Turning off one allows you to experience the other. Our sensors move through space, which exists in the present. Our visualizations move through time, which exists in the future outside of space. Visualization causes our sensors to turn off. When our sensors call us back to the present due to the sensing of some event which triggers them, our visualization goes away. ***The process of turning off our sensors to visualize, imagine or create a future event is the central theme of this book.***

Through visualizing going down a funnel, I found out that I could "tunnel out" of my body by turning off the sensors comprising it. This involves the ability to focus on the visualization that is occurring in my imagination. You can't just get a momentary glimpse of a funnel in your imagination and then find yourself out of your body. You have to stay in the funnel until you have traversed to and burst through its closed end. Having done so, you will find that your sensors have been turned off. This is not easy. Your sensors (eyes, ears, skin, etc.) constantly send signals to you which say, "Pay attention! I have something for you to experience!"

Leaving your body by turning off the sensors which comprise it is like leaving your house when you go on vacation. Before you leave, you check to make sure you have turned off all of the faucets, the lights, the television, the alarm clock in your bedroom, and so on. After you have turned all of those things off, you can leave your house and go on your vacation.

Leaving your body by turning it off is more easily done at night. At night, your sensors are at least partially turned off. You are in bed. Its dark, so your eyes are relatively inactive. It is quiet (hopefully) so your ears are relatively inactive. You are not eating, so your taste buds are inactive. And so on. Under these conditions, it is relatively easier to finish closing down your sensors and have an OBE.

On the other hand, during the day your sensors are very active. They are continuously receiving signals from the outside environment and sending them along to you. The world is a noisy place. Your ears are constantly active. So is your vision. You might be sensing smells in your environment or eating lunch or snacking. All of these incoming signals seek your conscious attention and make it very hard to ignore them by visualizing going through a funnel.

So the best time to have an OBE is at night when your consciousness is not receiving a lot of input from your sensors. Visualizing going through a funnel is much easier and you are more easily able to stay in the funnel until your sensors have been turned off so you can have an OBE. Once your sensors have stopped sending you signals, you may find that you have burst through the closed end of the funnel and are outside of your body.

Similarly, quite a few people have reported experiencing OBEs while they are under anesthesia while undergoing surgery. What does anesthesia do? It turns down your sensors so that you won't feel pain during an otherwise painful surgery or medical procedure.

We can't completely turn off our bodies. We can temporarily turn off the sensors that connect us to the real world (eyes, ears, skin, nose, tongue) but our bodies are comprised of more sensors and systems than these. For example, there are other sensors and systems that monitor our internal functions and organs that keep us alive such as our respiratory and cardiovascular systems. We can turn these systems down (breathing, heart rate, etc.) but we can't turn them completely off. We would die if we completely stopped breathing or if our hearts completely stopped pumping blood. In this book, when I equate our sensors to our bodies, I am referring to those sensors that we can turn off in order to have an OBE such as our eyes, our ears, our skin, our tongue and our nose. Turning our bodies entirely physically off is not necessary in order to have an OBE.

The reader will recall that earlier I said that I can't control when an OBE ends. This is because, even though my sensors are not sending any signals to my consciousness, they are still capable of doing so. Our sensors are always looking for signals to receive. Once they receive a signal, they send it along to our brains, including the parietal lobe, which passes their "construction" of the real world along to our consciousness. When we imagine going down a funnel, we are focusing on turning off the ability of our sensors to send us a signal. But if we are having an OBE and one of our sensors receives a strong signal (e.g. - an ambulance siren), the signal forces its way back into our consciousness and the OBE ends.

Returning to our house analogy, it is like having doors on our house. One of the purposes of a door is to keep people out of our house so that we may have privacy and safety in our own home. If our door is strong enough, a normal person won't be able to get in by breaking the door down. But if the person is very strong, he will be able to overcome the strength of our door and enter our house. Similarly, strong signals can "get in" to our consciousness and end an OBE.

We make the point that not all acts of turning off our sensors and visualizing a future event involve having an OBE. We constantly visualize, if only for a second or two. We all daydream. Daydreaming is a form of visualizing in which we imagine future events. Athletes visualize success before they compete. Have you ever seen an Olympic athlete (gymnast, downhill skier, etc.) on television just before they compete in an event? If you have, you may have witnessed them make a sort of blank stare as if they are looking into the future to see themselves be successful in the event they are about to undertake. That is exactly what they are doing. In fact, the very act of *thinking* about a future event involves visualization.

If we are able to visualize for a long enough period, our sensors will turn off and we will have an OBE. So visualization is a matter of degree. Daydreaming involves turning off our sensors for just a moment or two while we imagine some future event. Visualizing for a longer period (e.g. - tunneling) frees us from our sensors and we experience an OBE.

Takeaways

Let's summarize our OBE theory so far.
1. Our bodies are collections of sensors which relay signals, or events, from our environment to our brains which pass along their construction of the real world to our consciousnesses.
2. The information our sensors provide to us constrain our consciousnesses in space.
3. We use our imaginations to visualize events.
4. Some visualizations, such as thinking about something and daydreams, are of short duration.
5. Longer visualizations, such as visualizing going down a funnel until our sensors turn off, can lead to an OBE.
6. OBEs can end when one or more of our sensors become active and send signals to our consciousness.

CHAPTER 3

IMAGINING THE FUTURE

As we have discussed, our body is made up of sensors which create space by sending signals to our brains which process them and send them along to our consciousnesses. These signals are in the form of frequencies. Frequency can be described as the number of times something happens per second. Frequencies are commonly measured in Hertz (Hz). One Hertz is one time per second. Visible light is made up of a spectrum which vibrate at different frequencies. The frequency or frequencies that our eyes receive, processed by our brains and sent along to our consciousnesses, determine what we "see". Similarly, our ears hear different frequencies, our skin feels different frequencies (e.g. - heat radiation, wet, cold), our noses smell different frequencies, and so on. These different frequencies combine in our brains as brainwaves. Brainwaves, which can reach 100 Hz or higher, are classified in several different ranges: Delta (.5 - 4 Hz), Theta (4 - 8 Hz), Alpha (8 - 12 Hz), Beta (12 - 35 Hz) and Gamma (35 Hz and higher). There are an infinite number of brainwave frequencies

We recall from the above that there is a tradeoff between our sensors and our ability to visualize with our imaginations. If our sensors are active, they are sending signals to our consciousnesses which makes it harder to visualize. On the other hand, if our sensors are inactive, as at night when we are sleeping or almost asleep, our ability to visualize increases accordingly because our consciousnesses are not being bombarded with sensor input.

The more intense the signals being received by your brain at any given time, the higher the frequency of your brainwaves. For example, if you are at a concert with loud music being played, your brainwaves will be vibrating at a high frequency. On the other hand, if you are asleep, your brainwaves will be at a much lower frequency. In other words, when our sensors are active, our brainwave frequencies are relatively high. When our sensors are inactive, our brainwave frequencies decline.

There is experimental evidence for this. A scientific study has shown that, when the sensor input to our consciousnesses declines, the Gamma (35 Hz and higher) brainwave frequency contributions to our brainwaves decline and the theta (4-8 Hz) brainwave frequency contributions increase. (See, for example, "EEG Default Mode Network in the Human Brain: Spectral Regional Field Powers"; Chen, Feng, Zhao, Yin, and Wang; Center for Higher Brain Functions, Capital Medical University, Beijing, China; December, 2007.)

As mentioned earlier, there is a region in our brains called the "parietal lobe". The parietal lobe interprets the input from our sensors. It plays a role in our ability to judge size, shape, and distance. The parietal lobe also plays a role in functions such as navigation and controlling the body, image interpretation and understanding spatial orientation and direction. In other words, the parietal lobe is intimately involved in **taking the input from our sensors and creating the "space" that our consciousnesses experience**.

As we have seen, the act of visualizing turns down the sensor input to our consciousnesses. Focusing is the key. Focusing is the act of willfully controlling your sensors in order to turn their input down or (mostly) off. Most people can turn off their sensors for a second or two. Even the simple act of closing our eyes turns down our sensor input because our eyes are no longer receiving signals. We all daydream. Daydreaming is a form of visualization. But most people only daydream for a few seconds before their sensors pull them back to the "real" world. Then they turn off their sensors for a few more seconds to continue their daydream. If we are practiced at focusing, we may be able to visualize going down a funnel. **The longer we can keep going down the funnel, the longer our sensors are turned off.** And if we can stay in the funnel long enough until we get to its end, our parietal lobe will run out of "raw material" to "construct" the "bodies" that keep us in the present and we can experience an OBE.

We have been talking about our sensors creating frequencies which are processed by our brains (including our parietal lobes) into brainwaves which our consciousness perceive as "space". But what is "space"? Space is a collection of "events" we perceive as a result of our sensors being activated by external signals. Saying "hello" to a friend would be an event. Listening to music would be an event. Playing hide-and-seek with your children would be an event. An event is anything that is triggered by our sensors, processed by our brains into brainwaves and experienced by our consciousnesses.

An event can be defined as anything that creates brainwaves in you. Events can be ordered by their durations, or how long they last. In general, the longer the duration of an event, the longer it takes to tunnel out of it by turning off our sensors. This is because longer events tend to be more complex. For example, they may include the repeated activation of a multitude of sensors by multiple participants for the duration of the event. It takes a lot of focus to "turn off" an event such as this. Our bodies constantly experience the events that our sensors provide to us through brainwave frequencies. Events are packages of one or more brainwaves. The most elemental unit of an event is one brainwave.

The definition of the word, "future" is something that doesn't occur in the present. We can't access it with our sensors. We have to imagine it. As discussed in the Introduction, the future is imaginary time. It is hard to believe that we could exist without the ability to access the future by turning off our sensors. We need it to think, to plan, to daydream. The future is where OBEs occur.

It is like there is a door separating the present from the future. On one side of the door is the present where we receive input from our sensors that results in our consciousnesses experiencing events. On the other side of the door is imaginary time, or the future, where we can use our imaginations to visualize, or create, events that we would like to occur. The rooms on both sides of the door exist. But mostly, we stay on the present side of the door where our sensors are active. When we travel through the door to the future, the visits are relatively short. I believe this may be an evolutionary effect to protect us. When we are on the future side of the door, our sensors (eyes, ears, skin, nose, tongue) are turned off or way down. That places us in a certain amount of danger from environmental and other hazards. So our brains have developed to let us take relatively "quick visits" to the future. Have you ever all of a sudden got the feeling that something unexpected was about to happen and then it did? If you have, you have been to the future side of the door.

Overcoming the tendency of the visits to the future side of the door being short allows us to stay in the future and experience OBEs. As mentioned earlier, this involves developing the focus and concentration that will allow us to keep the door open. Everything can be learned. We can "train" our sensors to let us stay in the future for more than a fleeting moment. The process of learning is training our brains to let us accomplish what we seek. Doing back flips, playing an instrument, flying an airplane and a myriad of other accomplishments result from causing our brains to change. OBEs are no different. We can train our brains to allow us to focus and concentrate long enough to travel to the future where OBEs can occur. I do this by imagining going down a funnel.

Takeaways
1. Our sensors are activated by external signals.
2. The signals from our sensors are processed by our brains into brainwaves.
3. A processing center for sensory input is the parietal lobe.
4. An output of the parietal lobe is brainwaves that are perceived by our consciousnesses as events occurring in space.
5. An event can be defined as anything that creates brainwaves in you.
6. Events can be ordered by their durations.
8. Visualizing going down an imaginary funnel shuts off our sensors.
9. When our brain, including our parietal lobe, stops receiving sensory input, our bodies "let go" of us and we can experience an OBE.

CHAPTER 4
ANATOMY OF AN EVENT

Let's trace an ordinary, everyday, type of event from its beginning in the future to its sensing by us in the present. Note that the following discussion relates to the general case of events which come to us automatically through our sensors on a daily basis. The next chapter, Chapter 5, will deal with the special case of OBE's, which do not involve input from our sensors. Figure 1 below shows the event-time graph template that we shall use to describe events.

Figure 1

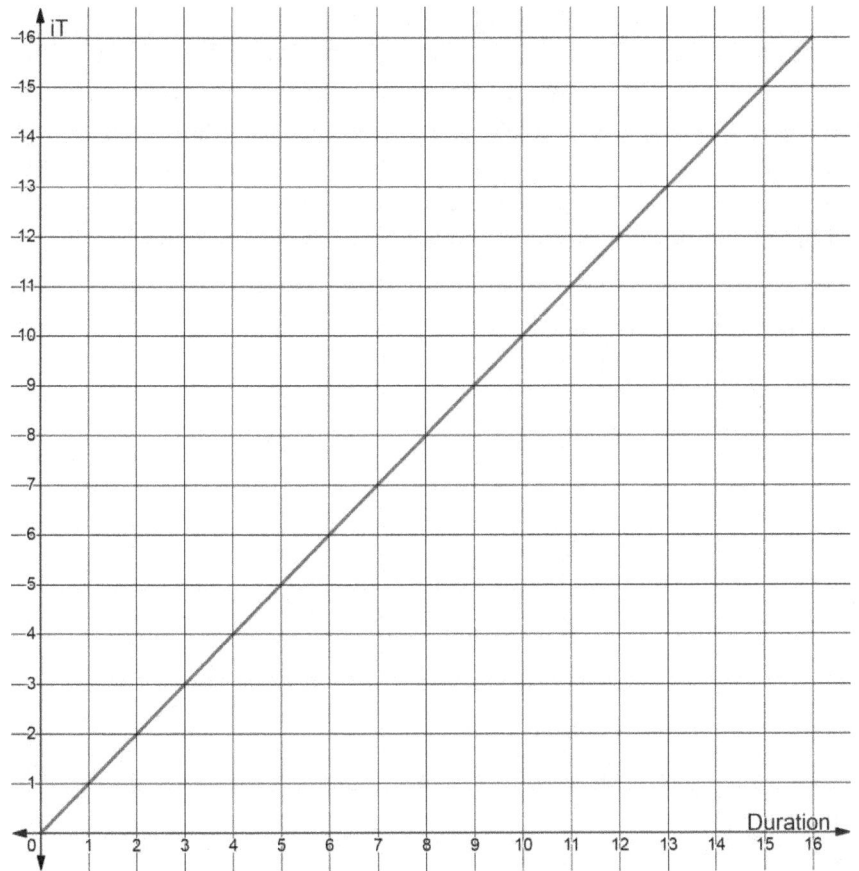

In Figure 1, the horizontal axis contains events, categorized in terms of duration, which is measured in the chosen time units (seconds, minutes, hours, etc.) that fit the event or events under consideration. The greater the duration of an event, the farther it is away on the horizontal events axis from the origin of the graph, or point (0, 0). If there is no event, the duration is "0". The horizontal axis of Figure 1 represents the present, or time "0". Events, or more precisely, the part of events that have impacted our consciousnesses through our sensors, are experienced by us in the present along the horizontal events axis.

The vertical axis in Figure 1 represents the future, or the time remaining until a particular event ends. The vertical axis is imaginary time which is denoted by the "i" (for "imaginary") in front of the "T". Time is imaginary because, until an event is completely over, we don't know how it will end. The unfinished part of the event remains in the future until the very end of it has impacted our consciousnesses through our sensors in the present at time "0". The future is imaginary because we can only imagine how it will unfold as it impacts our sensors.

Events flow through time. We know this because we experience a constant series of sensor-activating events throughout any given second, minute, hour, day, week, month, year, and so on. This means that the vertical time coordinate of any event on an event-time graph gradually decreases to "0", which is the time the event is experienced in the present. When the end time of an event has counted down to "0", the event will have started out in the future and trickled down to the present events axis where our sensors picked it up and passed it along to our consciousnesses. The idea of an event flowing through time until it reaches the present along the horizontal events axis is illustrated in Figure 2:

Figure 2

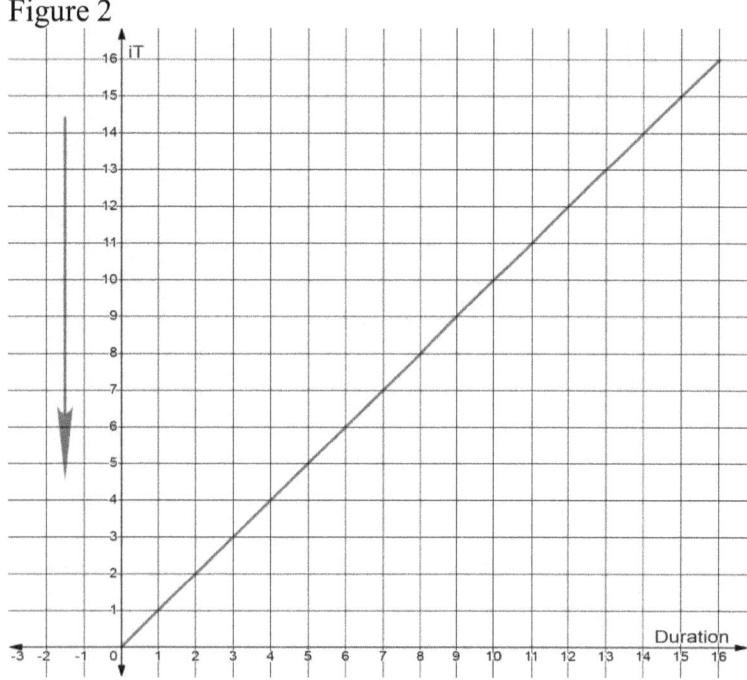

Figure 2 illustrates the fact that any event that begins at a specific time in the future will "trickle down" through time and eventually cross the horizontal events axis, or the present, where it will be perceived by us.

Just what do we mean by "trickle down"? Imagine a horizontal escalator. You can find horizontal escalators at many large airports. Horizontal escalators don't move you from one floor to another as department store escalators do. The word "horizontal" in "horizontal escalators" implies what they do. They move you horizontally from one location to another. Large airports use them because there are often large distances between the boarding and parking areas, for example, that some people have difficulty traversing.

Now consider an imaginary horizontal escalator that moves from the future to the present, a "time escalator". Instead of being measured in feet like the airport escalators, the length of the time escalator is measured in time. Suppose someone who is in the present is able to travel 10 hours into the future. That person puts a battery-operated video player and a note that says, "Push the red button.", in a box and then seals the box. The person, who is 10 hours in the future, places the box on the time escalator and away it goes towards the present.

As time ticks down, the box gets closer and closer to the present (time "0"). It starts out 10 hours away from the present. An hour later, it is 9 hours away from the present. And so on. Finally, the box "pops out" at some location in the present. A person who happens to be standing there picks up the box and opens it. He sees the note, pushes the red button and watches a video play.

The person who travelled to the future created an event 10 hours in the future. The event (box, instructions, video player, video) "trickled down" to the present and was experienced by the person who found and opened the box. The concept of time "trickling down" and the time escalator example will be used throughout this book. Time is a conveyor belt from the future to the present. It is imaginary because we can only imagine what the future will bring.

The Figures below provide an example of how an event of a specific duration and end time starts out in the future and ends up impacting our consciousnesses through our sensors in the present.

Figure 3

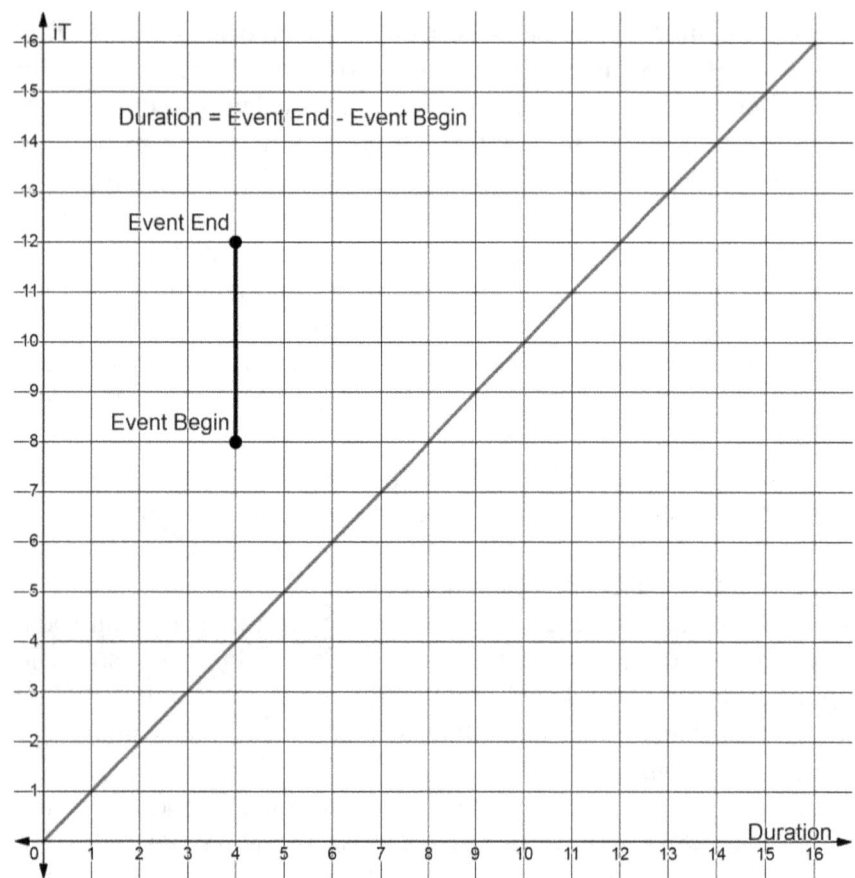

Figure 3 illustrates an event that starts out at point (4, 12) in the future, which means that the event will end in 12 time units and has a duration of 4 time units. (For the sake of this example, we will assume that the chosen time units are seconds.) Because the duration is 4 seconds, the event will begin to impact our consciousness along the horizontal event axis, the present, in 8 seconds (12 seconds - 4 seconds), as seen on the graph.

As time counts down, the event begin time and end time will count down with it. Figure 4 illustrates the situation after 3 seconds have passed:

Figure 4

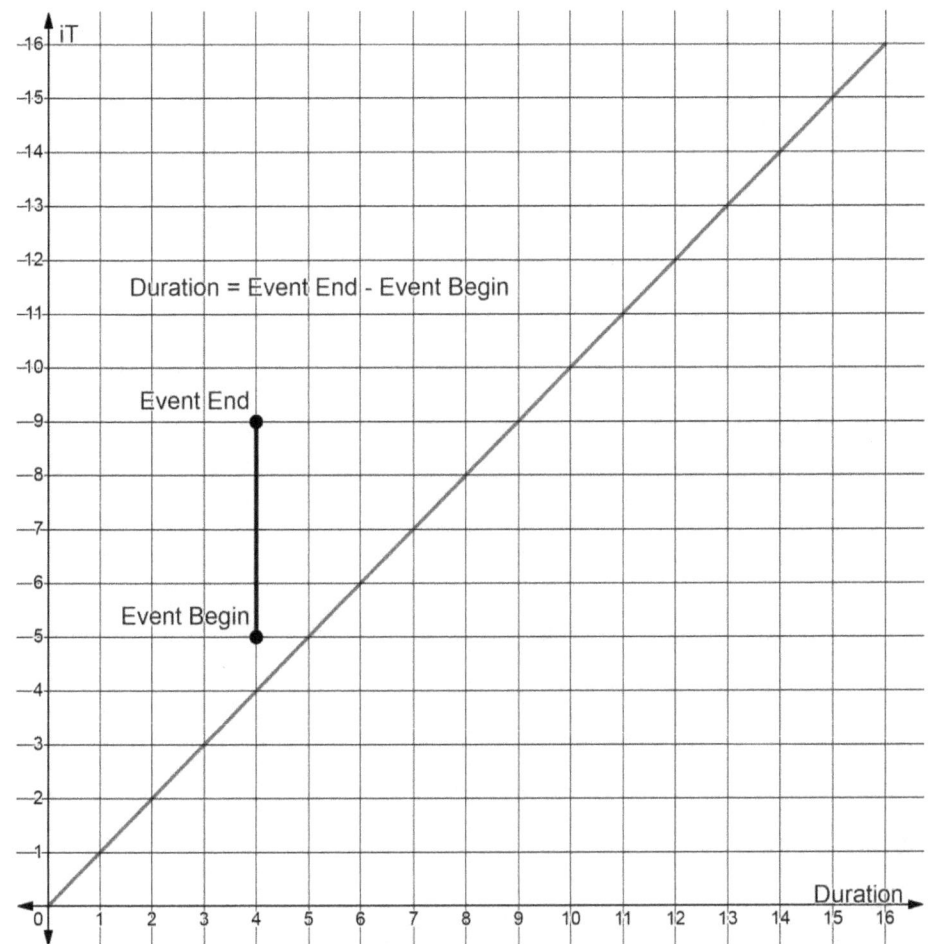

Figure 4 shows that, after 3 seconds have passed, the event is 5 seconds from beginning on the horizontal events axis (8 seconds - 3 seconds) and 9 seconds from ending on the horizontal events axis (12 seconds - 3 seconds). The duration of the event has remained at 4 seconds (9 seconds - 5 seconds). As time has counted down by 3 seconds, the entire event (beginning and end) has gotten 3 seconds closer to the present.

Figure 5 illustrates the scenario after 5 more seconds have passed:

Figure 5

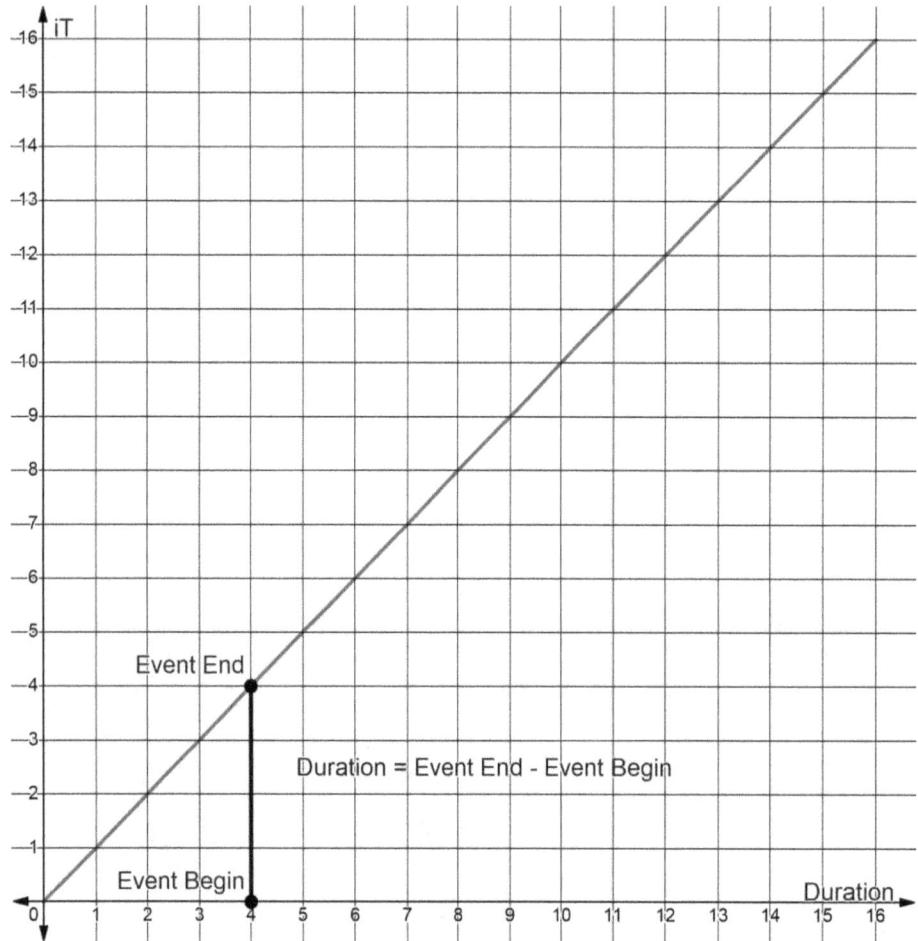

Figure 5 shows that, as time has counted down by 5 more seconds, the event is beginning to impact our consciousness on the horizontal events axis and is 4 seconds from ending on the horizontal events axis (9 seconds - 5 seconds). The duration of the event has remained at 4 seconds (4 seconds - 0 seconds).

In Figure 5, the event end lies on the diagonal that goes from point (0, 0) in the lower left-hand corner of the graph to point (16, 16) in the upper right-hand corner of the graph. The characteristic of the diagonal is that every point on it has the same value for that point's duration and end time (e.g. - (1, 1), (2, 2), (10, 10), etc.). This means that, for every point on the diagonal, the event will end after its duration has passed. This makes sense. After the event begins on the horizontal events axis, it will be over in whatever duration it has.

Figure 6 illustrates the event when it has moved below the diagonal:

Figure 6

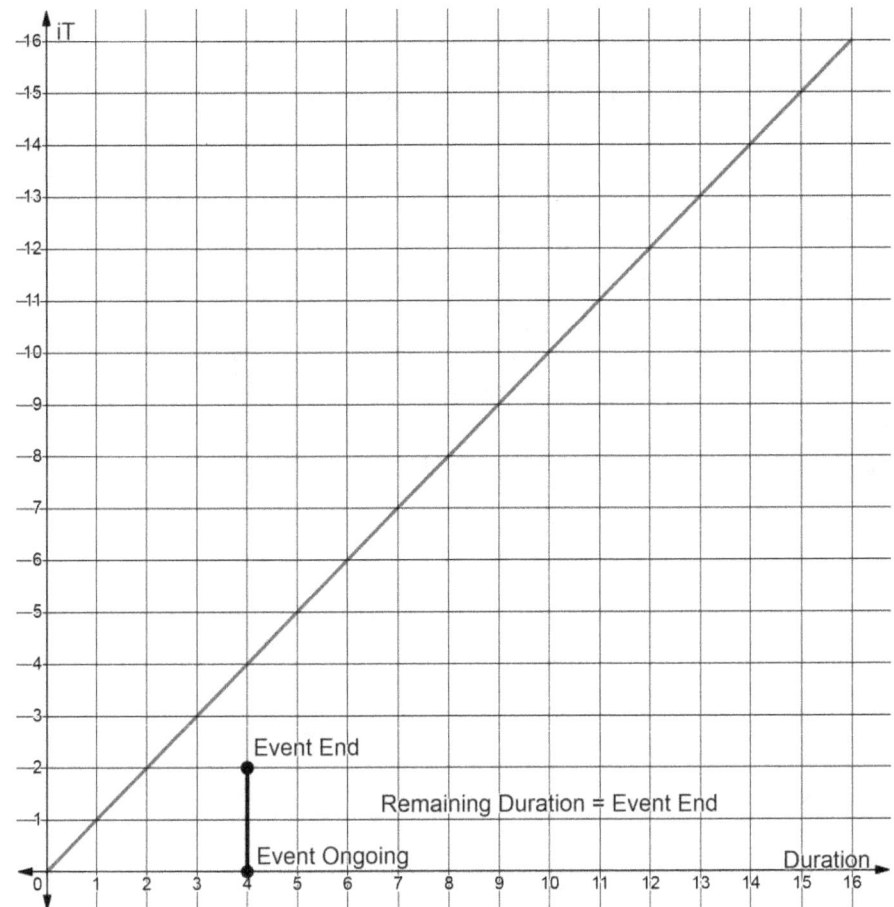

Figure 6 shows that the event will end in two seconds. The event is ongoing and has been received by our sensors for two seconds (original duration, 4 seconds, minus event end, 2 seconds). The event's remaining duration is 2 seconds. The 2 second part of the event that has been consumed by our consciousnesses occurred in the present along the horizontal events axis. The 2 second part of the event remaining above the horizontal events axis that has not reached our consciousness is still imaginary.

Imagine going to a baseball game. You have watched seven innings and play is about to begin at the top of the eighth inning. Each of the first seven innings was real to you because you experienced them in the present as they occurred. The two remaining innings are imaginary because they remain in the future. While the first seven innings are "in the book", the results of the next two innings are undetermined until they are played.

As events of various durations flow from the future to the present, our sensors pick them up and our consciousnesses bounce around along the horizontal events axis as we experience them. We usually "lock on" to a particular event and experience it in its entirety before our sensors select another event for us to witness. However, sometimes events we are experiencing get interrupted by other events that "take over" our sensors. Suppose you are sitting on a bench in a park eating ice cream, a very pleasant event. All of a sudden, an ambulance with its siren blaring comes down the street and passes by you. That event will, in all probability, consume your attention until the ambulance has passed.

The above example illustrated an event that began 12 seconds in the future and then trickled down to the present where our sensors picked it up and relayed it to our consciousnesses. Where did the event that began 12 seconds in the future come from? In Chapters 12 through 14 of this book, we will discuss how we can actually create events in the future that will be experienced in the present.

But first, we will address OBEs. OBEs are a special kind of event. We have seen that OBEs take place from beginning to end when our sensors are turned off. We have also seen that our sensors operate in the three-dimensional space of the present. Our sensors are what bring the present to our consciousnesses. Since OBEs don't involve our sensors, they don't take place in the present. They must therefore not only be created in the future, they must also be experienced in the future. Chapters 5 through 11 will deal with the creation and experiencing of OBEs. (Recall that Chapters 6 through 10 may be skipped by those who don't want a mathematical analysis of OBEs.)

CHAPTER 5

INSIDE EVENTS AND THE CREATION OF OUT-OF-BODY EXPERIENCES

We have seen that brainwaves are created when our sensors select an event for us to experience. Witnessing an event creates brainwaves that bring the event to our consciousnesses. But we don't have to witness an event to create brainwaves. Events can be classified as "outside" or "inside" events. "Outside" events are events picked up by our sensors and were the events discussed in Chapter 5. "Inside" events are events we initiate from within ourselves, our consciousnesses. Inside events create brainwaves just as outside events do.

Unlike an outside event that must trickle down to us in the present from the future, inside events are also experienced by us in the future a tiny fraction of a second after we create them. OBEs are inside events that occur after our sensors have been turned off. We turn off our sensors and move to the future where we visualize, or create, an inside event. The visualization creates brainwaves that flow back to our consciousnesses while we are still in the future. Our sensors are turned off during the entirety of an OBE. There are no active sensors during an OBE and therefore nothing to sense an OBE in the present, where our sensors exist. After we create the OBE inside event brainwaves in the future, we also experience the event in the future where the brainwaves created can flow back to our consciousnesses without the need for sensors. The fact that OBE brainwaves are generated, and experienced, in the future opens up a seemingly unlimited number of different OBEs that can be experienced. We can imagine literally anything in the future. We can fly through the air, move through walls, travel instantaneously to any location and so on.

When our sensors are active, they convey outside events to our consciousnesses. The outside events can only be perceived by us in the present after our sensors pick them up and relay them to our consciousnesses. When our sensors have been turned off, we have access to the world of the future where we can create events which do not require sensor input to experience.

So if we want to experience an "OBE", we need to turn off our sensors. When our sensors are turned off, we exchange outside events for inside events where OBEs occur. As mentioned above, I turn off my sensors by focusing going through an imaginary funnel. This allows me to ignore the demands on my consciousness made by my sensors while they are trying to convey outside events to me. If I can stay in the funnel long enough until I can exit its closed end, I will find myself in the future where I can create brainwaves by visualizing an event. After I create the brainwaves in the future, I experience them in the future while my sensors are still turned off.

In the case of inside events, including OBEs, we both create and experience the events while we are in the future. Contrast this to outside events which also originate in the future but are experienced in the present. We have postulated that outside events also originate in the future but we haven't yet discussed or theorized how they get there in the first place. We will consider this topic later on in this book.

The process of experiencing an OBE goes like this:

1. Focus on ignoring your sensors until they turn off.
2. Create brainwaves by visualizing an event while you are in the future.
3. Experience the brainwaves in the future a fraction of a second after you create them.

As I have said, I turn off my sensors by imagining going through an imaginary funnel. Focusing on going through an imaginary funnel works for me, and it is theoretically predicted as we shall see in Chapter 9, but others may focus on something else. The key is focusing on something that allows you to ignore your sensors until they turn off, or until your brain, including your parietal lobe, has no sensor input with which to construct the present and "lets go" of you. With respect to those who may do it differently than me, for the sake of consistency, I will refer to the focusing as I experience it: going through an imaginary funnel.

In a very real sense, I am tunneling out of my sensors when I focus going through an imaginary funnel. As mentioned above, your sensors comprise only part of your body. However, even though it is not entirely accurate, it is not hard to consider your sensors as being your entire "body". The entire outside of your body is composed of sensors, including your ears, eyes, nose, tongue and skin. The outside of your body *is* a sensor. There is nothing on the outside of your body that isn't a sensor, something that is connected to the *inside* of your body by nerves. Even the hairs on the outside of your body are sensors.

When video cameras became widely available, I saw a very interesting demonstration. A person connected his video camera to a computer monitor. He then turned the camera and monitor on and pointed the camera at the monitor. The result was that the image that formed was not of the monitor or the camera. It was a beautiful spiral! This was due to what is called "feedback". The camera basically looked at itself and then looked at itself looking at itself and then looked at itself looking at itself looking at itself, and on and on, which created the spiral. Similarly, when we examine our bodies, we are essentially examining our sensors with our sensors (e.g. - looking at our eyes with our eyes, touching our skin with our skin, etc.). Does that generate a feedback loop that makes our bodies "appear" as they not really are? Maybe our bodies' really "look" like they do when our sensors are off and we are in the future.

Figures 1- 6 above were used in our discussion of how we receive outside events through our sensors. We discussed that future events trickle down through imaginary time on the time escalator until they impact our sensors in the present and become recognized by our consciousnesses. We assumed that events with a fixed duration were somehow created at a specific time in the future without discussing how this might take place. As noted, we will discuss our theory about how outside (sensor-driven) events are created in the future later on in this book.

But for now we are going to continue to deal with OBEs, which we know are inside events. Figure 7 is an example of an inside event:

Figure 7

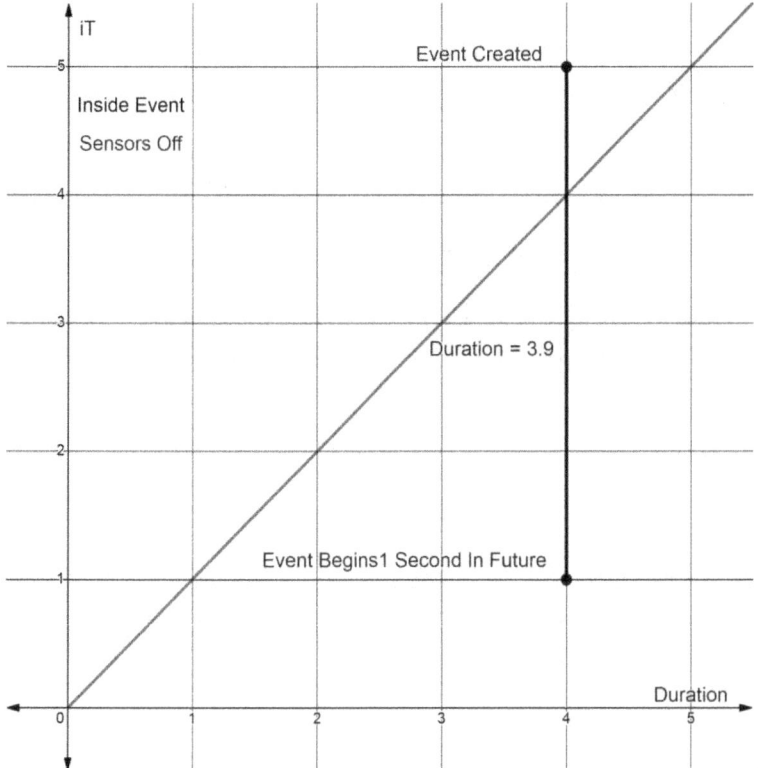

Figure 7 shows an inside event of duration 3.9 seconds. The event was created in the future ("iT") after we tunneled out of our sensors. We created the event when we were five seconds into the future. As we continued to add time to the event, the event portions already created were pushed down towards the origin. The difference between the events creation time (5 seconds into the future) and ending time (1second into the future) is 4 seconds. Then why is the duration listed at 3.9 seconds? The reason is because it will take us a small fraction of a second (in this example, one-tenth of a second) to begin experiencing the event after it has been created. You can think of this as lag time, or the time it takes us to "go" from five seconds into the future, where the event was created, to one second into the future, where we will experience the event. In other words:

    1.   ECT = ED + TTE + EBT

where:
    ECT = time event was created in the future
    ED = event duration
    TTE = time to experience the event after it has been created
    EBT = event begin time in the future

In our example:
    ECT = 5 seconds into future
    ED = 3.9 seconds
    TTE = .1 second
    EBT = 1 second into future
    So ECT = 3.9 seconds + .1 second + 1 second = 5 seconds

In the above example of an inside event OBE, our consciousnesses both created and received (experienced) the event. The entire event took place in the future. No sensors were needed.

We have discussed that events consist of brainwaves and inside events create their own brainwaves without the need for sensors. An event may consist of a single brainwave or many brainwaves that are different from one another. (We will later describe the variables that make brainwaves distinct.) Figure 8 below illustrates this:

Figure 8

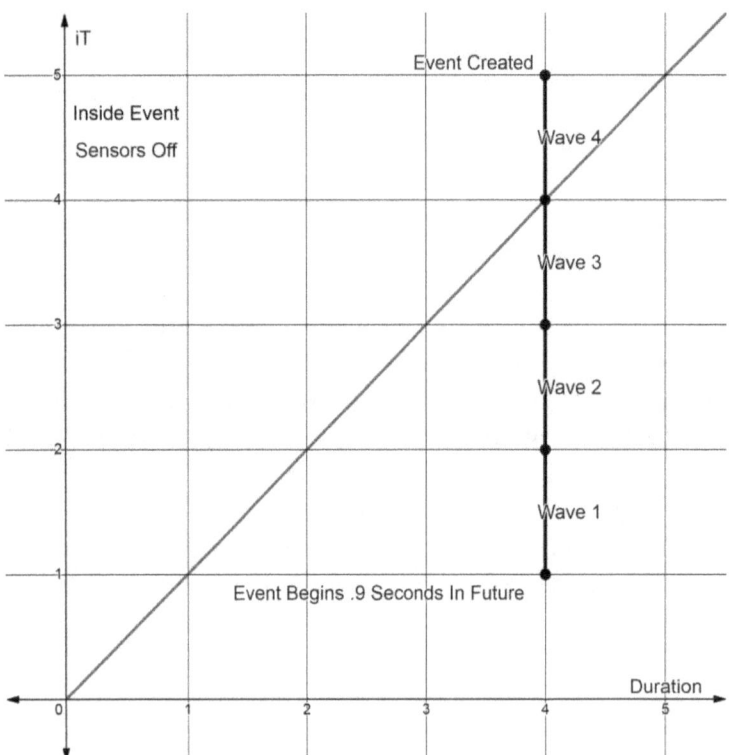

Figure 8 breaks up our inside OBE event into its four component parts: four brainwaves. The individual brainwaves exist between the dots. Wave 1 exists between seconds 1 and 2. Wave 2 exists between seconds 2 and 3. Wave 3 exists between seconds 3 and 4. And Wave 4 exists between seconds

4 and 5. As in Figure 7, we created the event consisting of four brainwaves when we were 5 seconds into the future. We first created brainwave 1, which is the first brainwave we will eventually experience as part of the event. We then created brainwave 2, which pushed brainwave 1 one second towards the present. After creating two more brainwaves, waves 3 and 4, we were done creating the event. As in Figure 7, we built in a delay of one-tenth of a second before we would start to experience the event (with brainwave 1) after it had been created. So the event begins .9 seconds into the future. Checking the math:

1.   ECT = ED + TTE + EBT

where:
ECT = 5 seconds into the future
ED = 4 seconds
TTE = .1 seconds
EBT = .9 seconds into the future
So ECT = 4 seconds + .1 seconds + .9 seconds = 5 seconds

Adding up the one-second duration of each of four waves gives a total duration of four seconds. Each of the waves in Figure 8 will in all probability have different characteristics because events evolve through time. After Wave 1 has passed through our consciousness, Wave 2 takes its place. Wave 3 eventually takes the place of Wave 2 and Wave 4 will take the place of Wave 3. We receive the waves in the same sequence they were created: 1, 2, 3, 4.

Inside events created in the future are received almost instantaneously. Figures 7 and 8 show that the inside event created in the future begins to be experienced one-tenth of a second after it is created. This is less than a blink of an eye (which takes approximately one-third of a second).

We can move around freely in the future precisely because it is imaginary. We are unbound and unhindered by our sensors. In the future we become waves. Waves have no mass, therefore no gravity to pin us to the present and slow us down. As has been mentioned, I believe our sensors are a cage that imprisons us in the present.

CHAPTER 6

CREATING WAVES WITH A CIRCLE

Gaining Further Insight Into OBEs

Thus far, we have developed our OBE theory without using mathematics. Using mathematics won't *prove* OBEs exist. But if we can find mathematics that help us *understand* OBEs, that will increase our confidence in our theory. We have discussed that inside events consist of self-generated brainwaves. We will be looking for an equation that "allows" brainwaves to be created in the future. But first we provide a tutorial on some basic concepts we will need to understand brainwaves. This material will involve basic Algebra and Trigonometry. Any reader who is not interested in this mathematical analysis or already is familiar with it may skip ahead to Chapter 12.

The Unit Circle

Figure 9 below illustrates what is called the "Unit Circle".

Figure 9

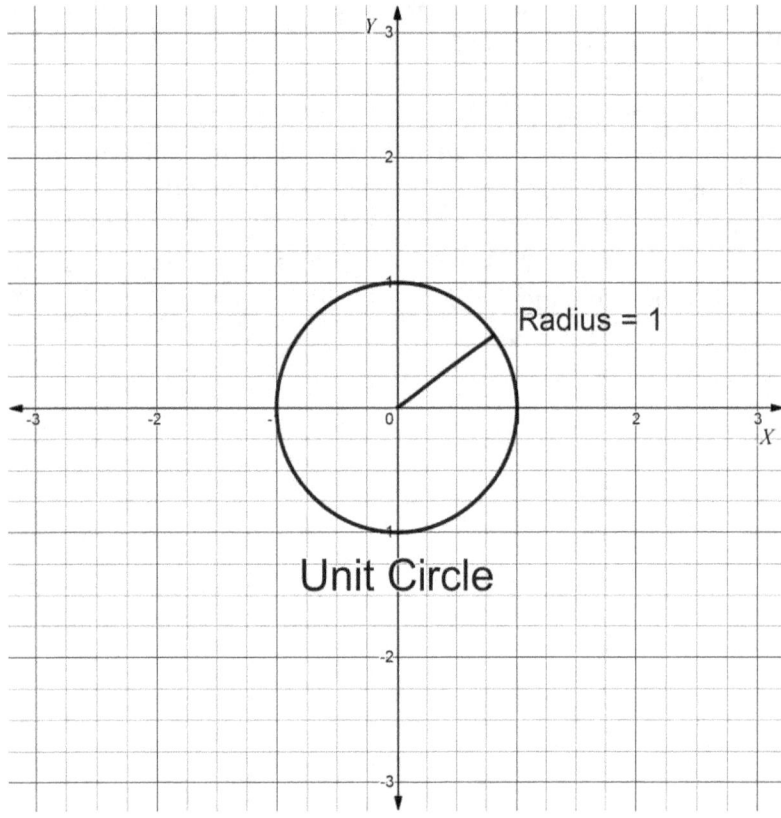

Figure 9 shows a coordinate system whose horizontal axis is "X" and whose vertical axis is "Y". A circle is centered on the graph at point (0, 0). The radius of the circle is 1, as seen from the graph. The reader will recall from his or her Geometry class that the distance around any circle, or its circumference, is given by:

2. $C = 2\pi r$

where C is the circumference of the circle of radius r and π is the familiar constant that represents the ratio of any circle's circumference to its diameter.

When the radius, "r", is one "unit" of any measure (inches, feet, meters, etc.), or "1", the circumference of its circle is:

3. $C = 2\pi*1$, or 2π

and the circle is called a "unit circle". (Note that the symbol "*" means multiplication.)

Now let's look at right triangles. Figure 10 below shows a right triangle:

Figure 10

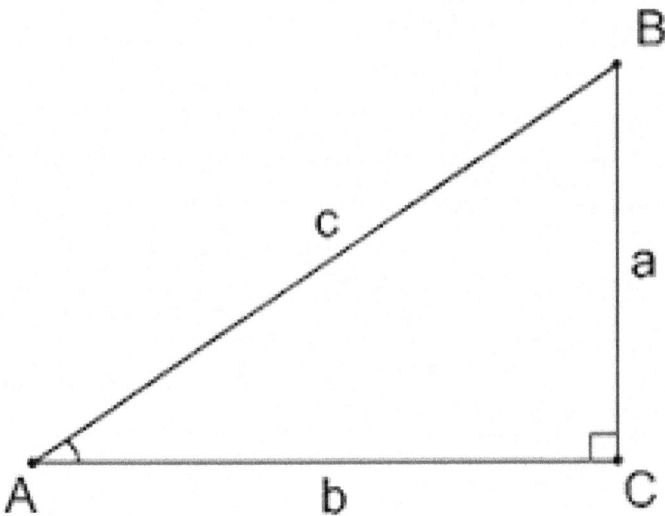

The triangle in Figure 10 is called a "right triangle" because two of its sides, AC and BC, form a 90-degree, or "right" angle when they intersect at point C. The right angle is denoted by the square drawn at point C. We live in a world of right angles. The walls of our houses (usually) form right angles with the floor and ceilings. The pages of the books we read have four corners that are right angles. Crosswalks are at right angles to the sidewalks that bound them. And so on. If two things (lines, walls, etc.) are at right angles to one another, they are said to be "orthogonal".

There are two other angles in Figure 10 besides the right angle at point C. The curved line drawn at point A represents the angle between sides AC and AB that terminate at point A. This angle is referred to as "angle A". There is also an angle B between sides BA and BC that terminate at point B that is referred to as "angle B".

The sides of a right triangle enjoy mathematical relationships with its angles, including "Sine" and "Cosine". The Sine of a right triangle angle is defined as:

4.   Sine of angle x = opposite side/hypotenuse

where "opposite side" is the side of the right triangle directly opposite angle x and "hypotenuse" is the longest side of the right triangle. (Note that the symbol "/" means division.) The Sine of angle A in Figure 10 is therefore:

5.   Sine(A) = a/c.

The Cosine of a right triangle angle is defined as:

6.   Cosine of angle x = adjacent side/hypotenuse

where "adjacent side" is the non-hypotenuse side of the right triangle directly adjacent to angle x. The Cosine of angle A in Figure 10 is therefore:

7.   Cosine(A) = b/c.

Note that, if the hypotenuse of a right triangle is "1", then the Sine of angle x is the value of its opposite side and the Cosine of angle x is the value of its adjacent side. In the case of Figure 10, assuming c = 1:

8.   Sine(A) = a/1 = a.

9.   Cosine(A) = b/1 = b.

Now we return to the unit circle.

Figure 11

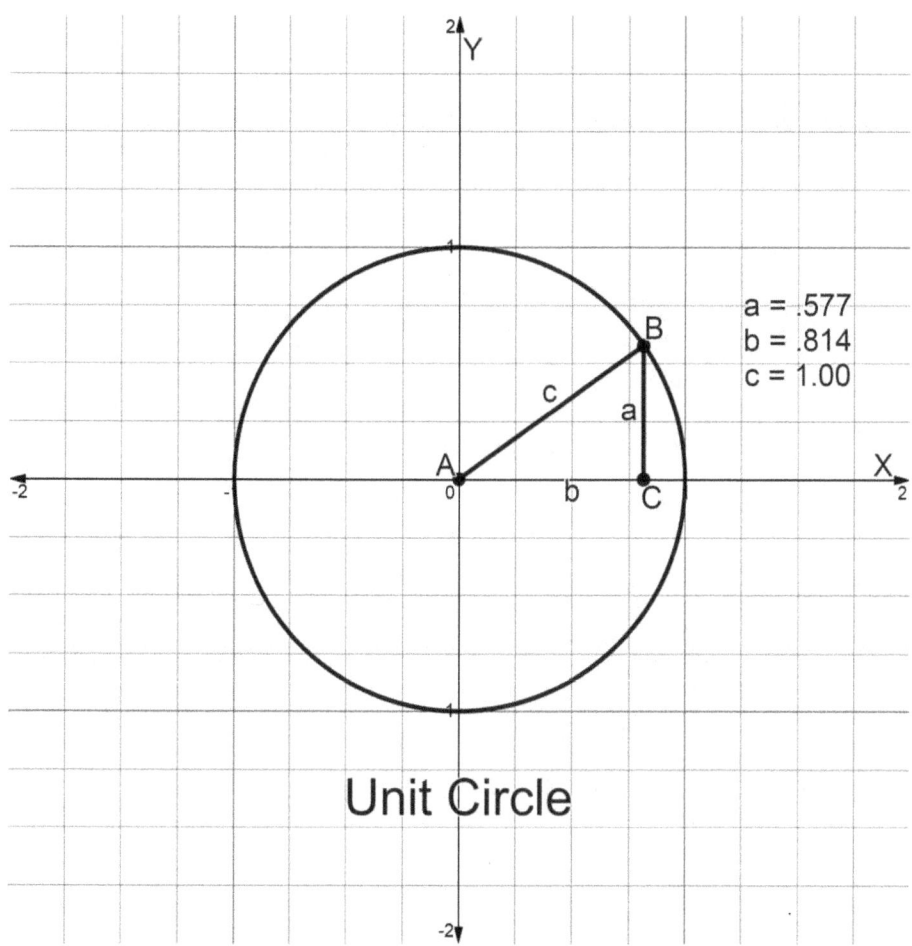

Figure 11 repeats the unit circle shown in Figure 9, with some additions. A right triangle has been inscribed inside the unit circle. Line AB is the radius of the unit circle whose length is "c" or 1unit (inches, feet, yards, etc.). If this weren't the case, the circle would not be a unit circle. Point B, a corner of the triangle, rests on the circle and its coordinates on the graph are (b, a). In other words, "b" is the horizontal "x" coordinate of point B and "a" is the vertical "y" coordinate of point B. From point B, we draw a vertical line to point C, line BC. The length of line BC is "a", or .577 units. Line AC is at the bottom of the right triangle with a length of "b", or .814 units, as seen from Figure 11. Angle A is the angle formed between sides AB and AC, angle B is the angle formed between sides BA and BC and Angle C is the angle formed between sides CB and CA. Because angle A is located at the center of the unit circle, angle A is referred to as a "central angle".

Angle C is a 90-degree angle because side BC is perpendicular to side AC and triangle ABC is therefore a right triangle. Because of this, we can determine the Sine and Cosine of angle A using the formulas from above:

10. Sine(A) = opposite side/hypotenuse = a/c = .577/1 = .577.

11. Cosine(A) = adjacent side/hypotenuse = b/c = .814/1 = .814.

Given the above, we can see that the x and y coordinates of any point on a unit circle represent the Cosine and Sine of the central angle formed by drawing a line (radius) from that point to the center of the circle.

Now let the radius of our circle increase to 2 units as seen in Figure 12:

Figure 12

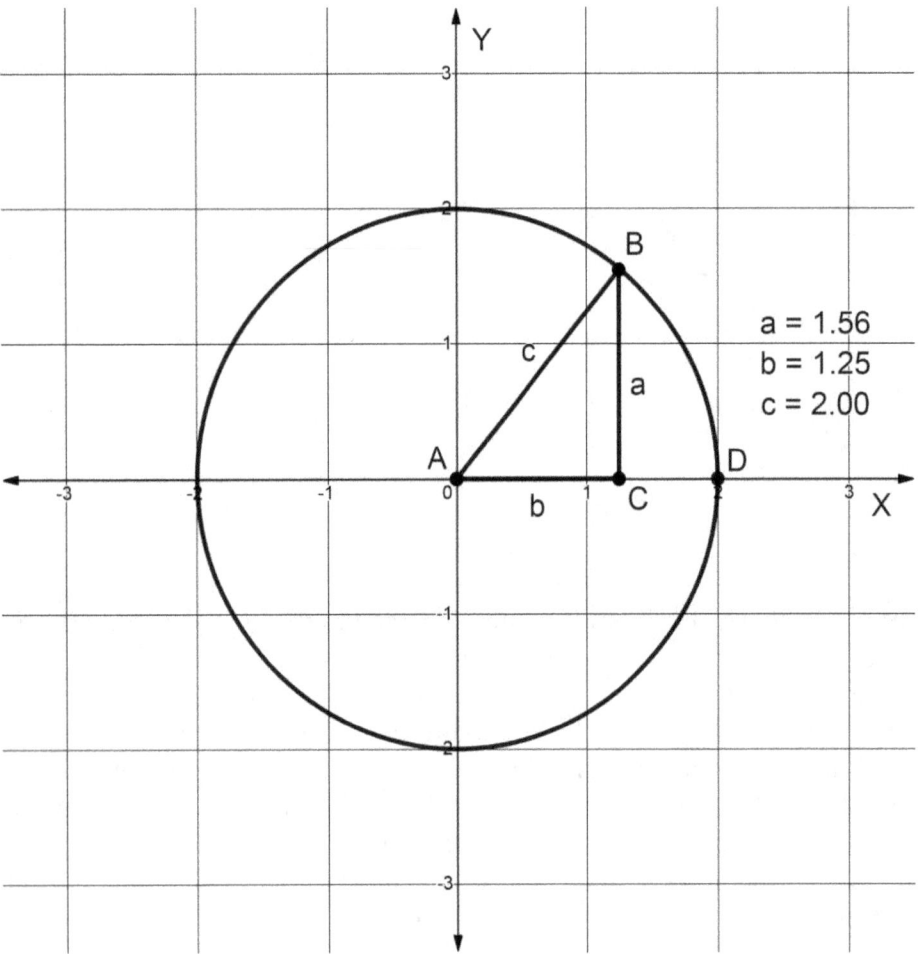

The circle in Figure 12 has a radius of 2 units ("c"), so it is no longer a unit circle. (If the radius of a circle is less than 1 unit or greater than 1 unit, it is not a unit circle.)

The Cosine of central angle A is no longer the horizontal component of point B and the Sine of central angle A is no longer the vertical component of point B because the radius, or hypotenuse, of the circle is greater than 1. We saw above that:

12. Sine(A) = opposite side/hypotenuse = a/c

so Sine(A) = 1.56/2.00 = .78, and

13. Cosine(A) = adjacent side/hypotenuse = b/c

so Cosine(A) = .125/2 = .625.

For a right angle inscribed in a circle, the hypotenuse of the right angle is always the radius of the circle, r. So we can redefine the equations for the Sine and Cosine of a central angle ("A") as:

14. Sine(A) = opposite side/r

15. Cosine(A) = adjacent side/r

As seen in figures 11 and 12, the adjacent side of a central angle is equal to the horizontal component of point B, or "x", and the opposite side of a central angle is equal to the vertical component of point B, or "y", regardless of the size of the radius. Substituting again:

16. Cosine(A) = x/r

17. Sine(A) = y/r

Solving equations 16 and 17 for "x" and "y", we get:

18. x = r Cosine(A) and

19. y = r Sine(A)

"Cosine" is usually abbreviated "cos" and "Sine" is usually abbreviated "sin", so equations 18 and 19 become:

20. x = r cos(A) and

21. y = r sin(A)

Equations 20 and 21 are the general equations for finding the horizontal and vertical components of any point on a circle (point "B" in figures 11 and 12) that is connected to the circle's origin by the circle's radius, "r", forming a central angle, "A", with the origin.

The base of the central angle formed by drawing a radius from a point on the unit circle to the center of the circle will always be the horizontal x-axis. As any point located on a unit circle moves around the circle, its horizontal "x" coordinate and its vertical "y" coordinate, or the Cosine of the central angle and the Sine of the central angle formed by a line from the point to the center of the circle will change.

There is a one-to-one correspondence between a point lying on the unit circle to the central angle formed by drawing a line (radius) from that point to the center of the circle. In other words, for example, as point B in Figure 12 moves around the circle, the central angle formed by drawing a radius from point B to the center of the circle will change proportionately. Because of this, the length of the segment of the unit circle from where the unit circle intercepts the x-axis to a point on the unit circle is also referred to as an angle.

In Figure 12, an example of this length is the length of segment (or "arc") DB. So the angle created by point B's position on the unit circle can be referred to as central angle A *or* the length of arc DB. Central angles, are usually measured in degrees. We know from our Geometry class that there are 360 degrees in a circle. "Arc angles", or angles based on arc length, are measured in radians. A radian is a measure of how many of a circle's radiuses can fit in the selected arc length.

How many radians are there in a circle? This is easily found by dividing the total arc length of a circle, or its circumference, by the length of that circle's radius:

22. Radians in a circle = circumference of circle/radius of circle.

Radians in a circle = $(2\pi r)/r = 2\pi$.

Equation 22 shows that there are 2π radians in any circle because the size of a circle's radius cancels out in the calculation. Imagine having a large circle. Make a string the length of its radius and then lay it end-to-end around the circle. You will be able to do this 2π, or approximately 6.3, times. Now make the circle smaller and repeat the exercise. You will still be able to go around the (smaller) circle 2π times. This shows that the radian measure of circles is consistent regardless of the size of the circle.

Because there are 2π radians per circle and also 360 degrees per circle, we can see that 90 degrees is equivalent to $\pi/2$ radians, that 180 degrees is equivalent to π radians, that 270 degrees is equivalent to $3\pi/2$ radians, and so on, by dividing each measure by the same number.

The one-to-one correspondence between central angles and arc length angles allows us to measure Cosines and Sines of angles that are expressed in either degrees or radians. The angles in Equations 20 and 21 are measured in central angles ("A" in Figures 11 and 12). If we instead measure angles in radians, Equations 20 and 21 could be distinguished as:

23. $x = r \cos(R)$ and

24.   y = r sin(R)

where "R" is the radian measure of the angle we are measuring ("B" in Figures 11 and 12).

It is important to not mix up degrees and radians when we are calculating Sines and Cosines. Two degrees is not equal to two radians! Most calculators have a setting in which the user can switch between degrees and radians.

Figure 13 shows a circle of radius 2 with radians 0π, π/4, π/2, 3π/4, π, 5π4, 3π2, 7π4 and 2π marked off in a counterclockwise direction:

Figure 13

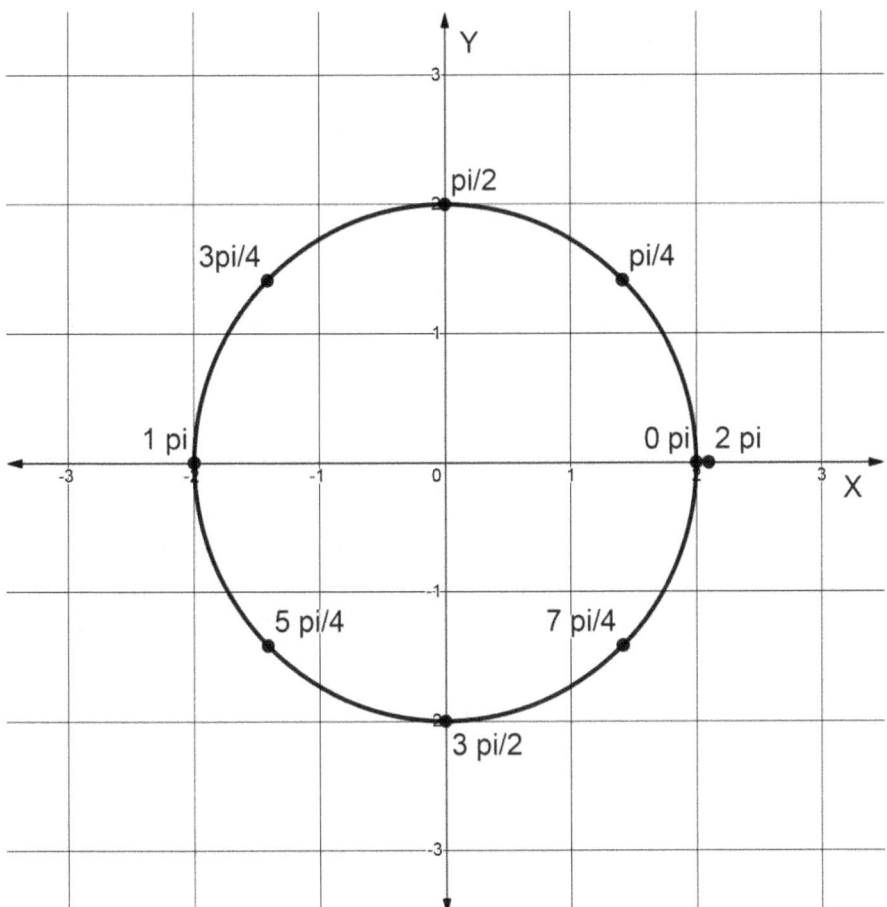

In Figure 13, the radian angle measures start out where the circle intersects the horizontal x-axis. The radian measure there is "0" because there is no arc at that point. From there, the radian measures move in a counterclockwise direction around the circle until they have returned to the intersection of the circle with the horizontal x-axis. At that point, the radian measures have moved completely around the circle for a distance of 2π radians.

Table 1 uses Equation 24 to graph the vertical y-axis coordinate of each point around the circle of Figure 13:

Table 1

Point in radians	Sine of point	*	Radius of circle	=	Y-axis component of point
0	0		2		0
π/4	.707		2		1.414
π/2	1		2		2
3π/4	.707		2		1.414
π	0		2		0
5π/4	-.707		2		-1.414
3π/2	-1		2		-2
7π/4	-.707		2		-1.414
2π	0		2		0

Table 1 shows that, as the radian measure increases from "0" to "π/2", the y-axis coordinate increases from "0" to a maximum value of "2". As the radian measure continues around the circle to "π", the y-axis coordinate declines from "2" to "0". From "π" to the bottom of the circle, "3π/2", the y-axis component decreases to "-2", which is the opposite of its value at the top of the circle. Finally, completing the 2π radian trip around the circle, the y-axis component finishes where it started at "0".

By changing the second column of Table 1 from "Sine of point" to "Cosine of point", we could calculate the X-axis component of that point by again multiplying column 2 by column 3. We leave this exercise to the reader.

Figure 14 graphs the radian measures of Table 1 and their Y-axis components (columns 1 and 4):

Figure 14

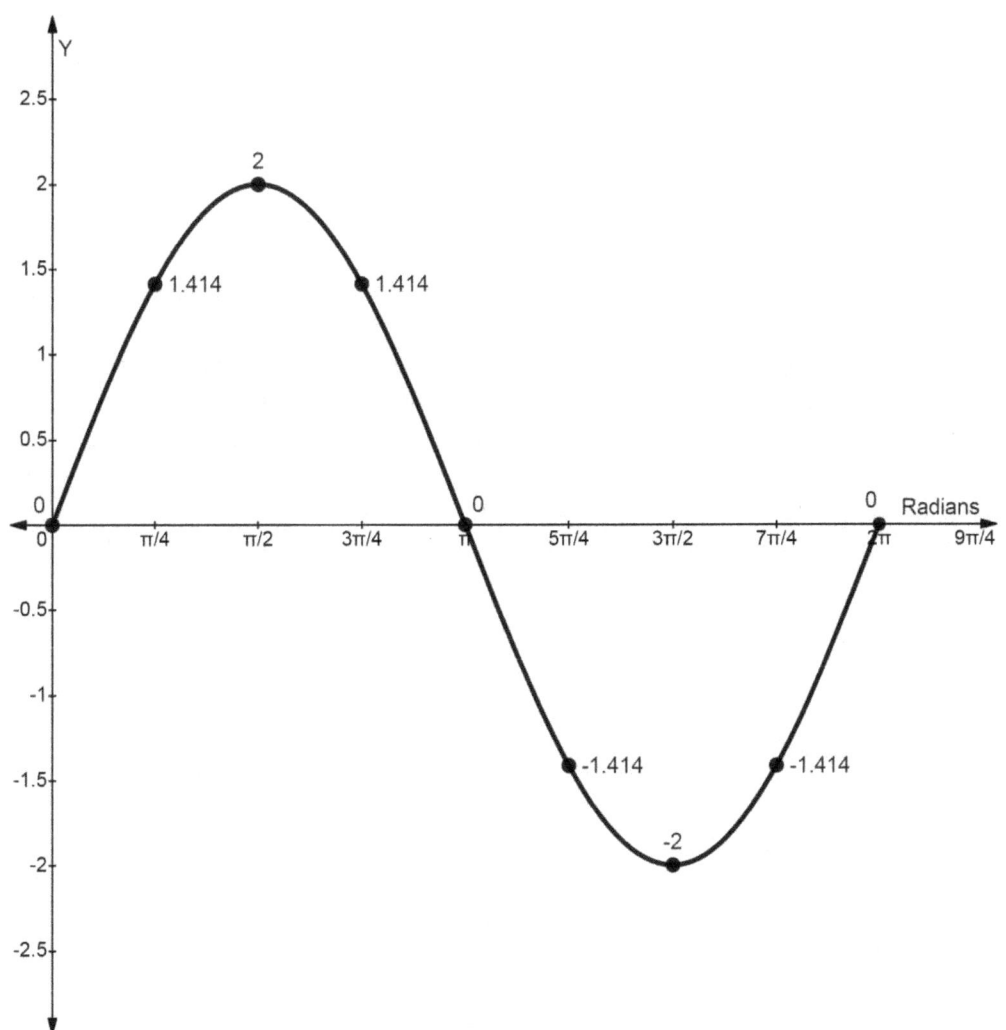

We can see from Figure 14 that, when the radian measure is "π/4", the y-axis value is 1.414, when the radian measure is "π/2", the y-axis value is 2, and so on. Figure 14 is a graphical picture of Table 1. It results from travelling counterclockwise around a circle of radius 2 units and recording the y-axis coordinate values of the radian points on the perimeter of the circle.

We travelled around the circle in Figure 13 one time which resulted in Figure 14. But there is nothing to prevent us from travelling around the circle multiple times in either direction. For example, when we completed one counterclockwise revolution around the circle in Figure 13, we moved from 0 radians at the beginning of our journey and completed the 2π radian trip around the circle which brought us back to where we started. But what if we make another round trip? In that case we will travel an additional 2π radians for a total of 4π radians. Every time we make a counterclockwise trip around the circle, we add an additional 2π radians to the distance we have travelled. Equation 25 below equates the radian distance we have gone to the number of trips we have made around the circle:

25. $RD = 2\pi n$

where "RD" is the radian distance around the circle (always 2π, as we have seen) times the number of trips around the circle, "n", we have made.

We can also move in the other direction, clockwise, around the circle in Figure 13. Nothing changes except the direction we have travelled. We distinguish the two directions by assigning positive values to counterclockwise travel and negative values to clockwise travel. In Equation 25 above, a positive "n" would denote counterclockwise travel and a negative "n" would denote clockwise travel.

We point out that the radian distance we have travelled is related to, but not equal to, the actual (physical) distance we have travelled around a circle. (This distinction is not necessary for the analysis herein but we digress for a moment in order to avoid any confusion.) We have seen above that the radian distance around any circle is 2π radians, regardless of the circle's radius. But the distance around a circle is its perimeter which is given by:

26. $P = 2\pi r$

where "P" is the perimeter of a circle and "r" is its radius.

We combine Equations 25 and 26 to get the actual distance we have travelled around a circle:

27. $AD = 2\pi n r$

where "AD" is the actual distance we have travelled around a circle, "r" is its radius in some selected unit (inches, feet, yards, etc.) and "n" is the number of revolutions around the circle we have made.

We can substitute Equation 25 into Equation 27 since the radial distance, "RD" equals "$2\pi n$":

28. $AD = RD\, r$

where, as before, "AD" is the actual distance we have travelled, "RD" is the radial distance we have travelled and "r" is the radius (in some selected unit) of the circle we are travelling around. (If the radial distance, RD, is negative, the actual distance, AD, will also be negative which means we have travelled clockwise around the circle.)

Having discussed the above, we present Figure 15, which extends Figure 14's graph in the positive direction as we move around the circle of radius r four times in the counterclockwise direction:

Figure 15

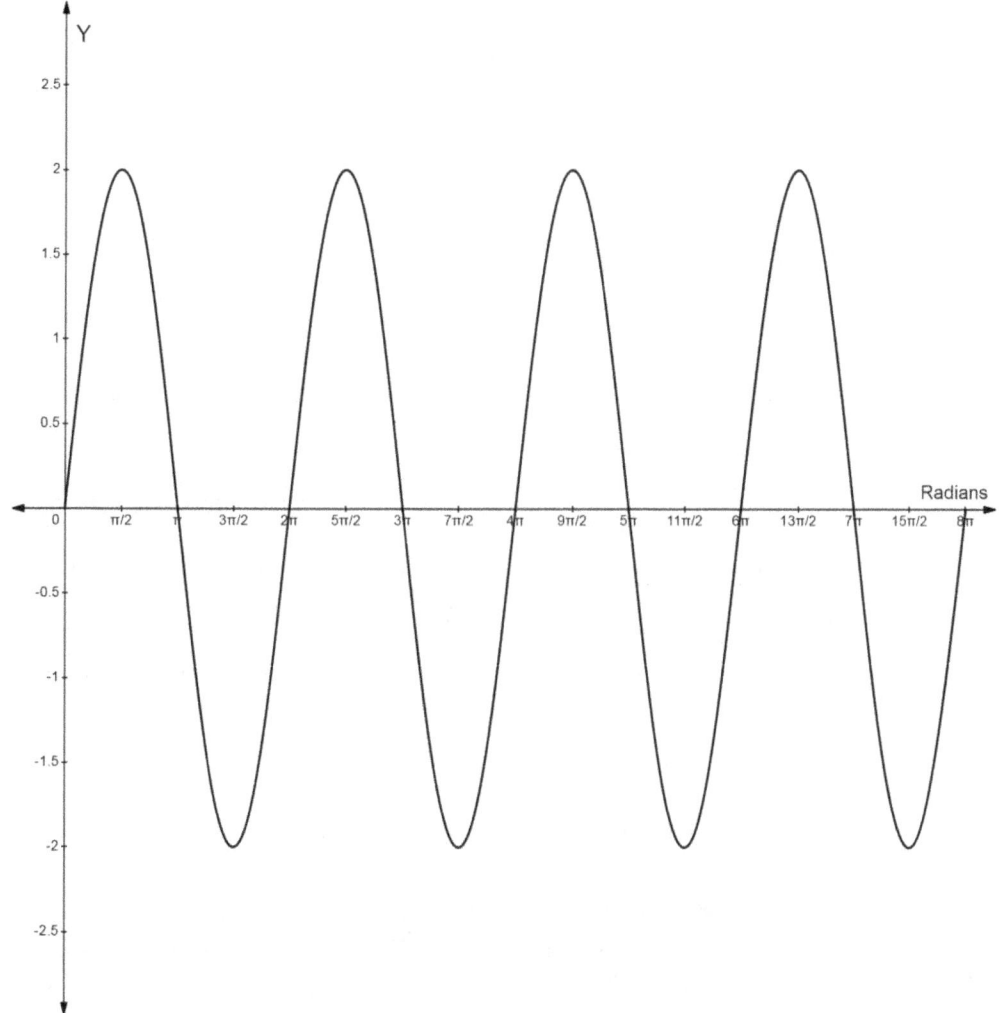

Figure 15 shows that we can create a wave by moving around a circle. The wave in Figure 15 was constructed from the y-axis coordinates of the circle as we moved around it, or y = 2sin(B), where "2" was the radius of the circle and "B" was the radian measure of the points around the circle (Equation 24). Because the wave in Figure 15 was constructed from the Sine function, we call the wave a "Sine wave". If we had instead used Equation 23 to measure the x-axis coordinates of the circle as we moved around it, or x = 2cos(B), we would call the wave a "Cosine wave".

Notice that the amplitude of the wave in Figure 15, or the vertical distance from the horizontal axis to the peak (or trough) values, is "2", or the radius of the circle that generated Figure 15. The radius of the circle that generates a Sine or Cosine wave will always be the amplitude of the wave.

We mentioned above that "we can create a wave by moving around a unit circle". This was useful for an understanding of the relationship between waves and the unit circle, but we can make it more precise. For example, what happens as we speed up how fast we go around the circle while we are creating a wave? It is easy to see that the peaks and troughs of the waves will be closer together. On the other hand, if we slow down how fast we are traversing the circle, the peaks and troughs of the waves will be farther apart. In other words, the frequency of the wave will increase or decrease depending on how fast we go around the circle. Equations 23 and 24 express the x-axis and y-axis coordinates of any point on the circumference of a circle as a function of the radian measure of the point. How do we incorporate frequency into the radian measure of points on the circumference of a circle as we move around it? Frequency in terms of going around a circle is expressed as the amount of time it takes to make one complete revolution around the circle. For example, starting out at point 0 in Figure 13 where the circle intersects the horizontal axis and then going all the way around the circle and returning to the starting point on the horizontal axis in one second would be a frequency of 1 Hz (1 cycle per second). Going around a circle twice in one second would be a frequency of 2 Hz, and so on. Suppose the frequency is 3 Hz. How many radians would we have traversed in two and one-half seconds? Consider Equation 29:

29. $B = 2\pi f\, t$

where "B" radians = (2π radians/cycle) * (f cycles/second) * t seconds. (Note that "radians/cycle" is read "radians *per* cycle" and "cycles/second" is read "cycles *per* second".)

Substituting our values into Equation 29:

$B = 2\pi$ radians/cycle * 3 cycles/second * 2.5 seconds = 15π radians.

("π" is usually not multiplied out as part of the calculation. Hence the radian value of "15π" given above.)

Since Equations 23 and 24 were expressed in terms of radians, and since Equation 29 above equates radians to frequency and time, we can substitute Equation 29 into Equations 23 and 24. The result will be that frequency and time will have been introduced into our wave equations (and graphs):

23. $x = r \cos(B)$

30. $x = r \cos(2\pi f\, t)$

24. $y = r \sin(B)$

31. $y = r \sin(2\pi f\, t)$

where:
"x" is the horizontal component of the point whose radian measure is "B",
"y" is the vertical component of the point whose radian measure is "B",

"r" is the radius of the circle,
"f" is the frequency in cycles per unit of time,
"t" is the duration of the wave in the same units of time as "f"

Equations 23 and 30 and Equations 24 and 31 tell us that radians can be measured in either arc length *or* time. The connection between the two methods is the frequency at which a point is moving around a circle.

By way of illustration, Figure 16 repeats Figure 15, but uses a frequency of .159 Hz [$1/(2\pi)$ Hz] between time 0 seconds and 25.1 seconds (the frequency of the wave in Figure 15):

Figure 16

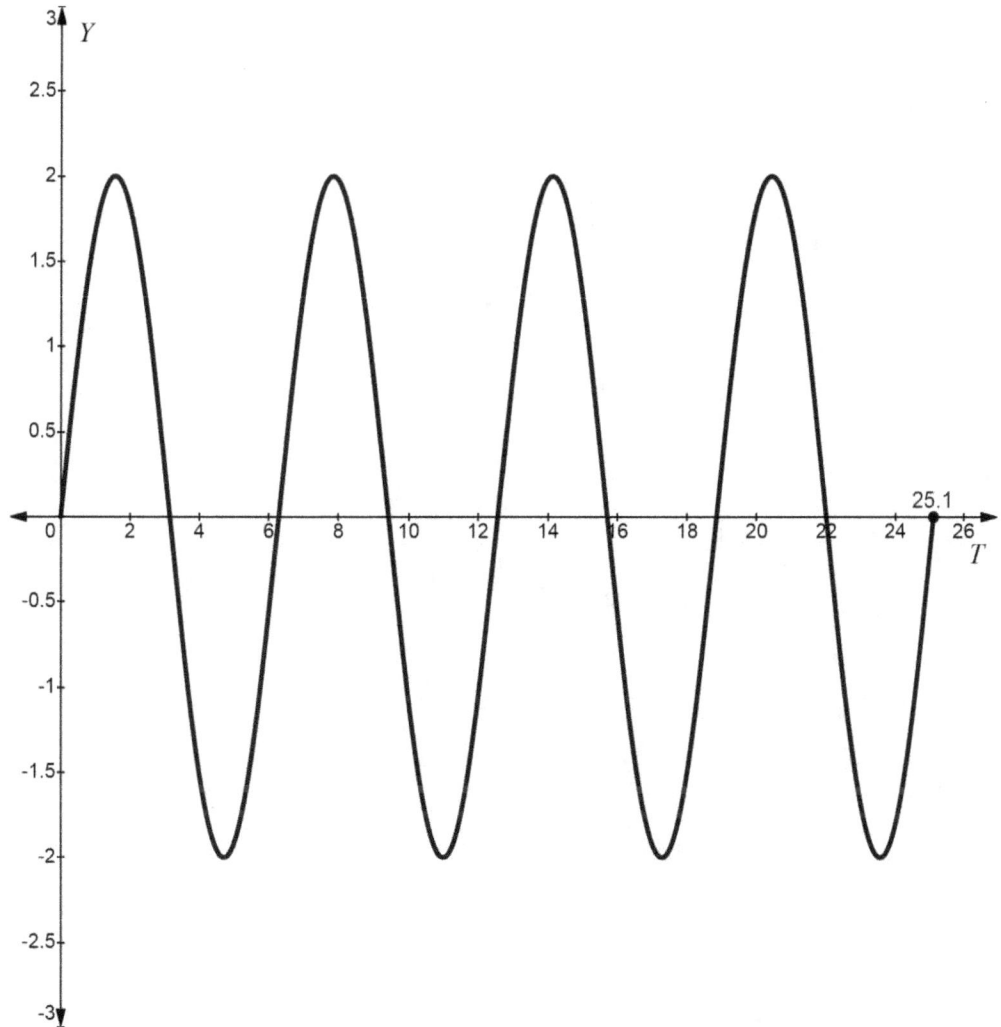

A comparison of Figures 15 and 16 shows that they are the same shape. Note that the x-axis in Figure 15 is measured in arc lengths (angles) whereas the x-axis in Figure 16 is measured in time (seconds). Radians measured in terms of angles and radians measured in terms of equivalent time units result in identical Sine waves.

Now that we have introduced frequency into Equation 31, we can easily change the frequency of the wave shown in Figure 16. Figure 17 has the same radius and time frame used in Figure 16 but now uses a frequency of .318 Hz (1/π Hz), twice the frequency of the Sine wave in Figure 16:

Figure 17

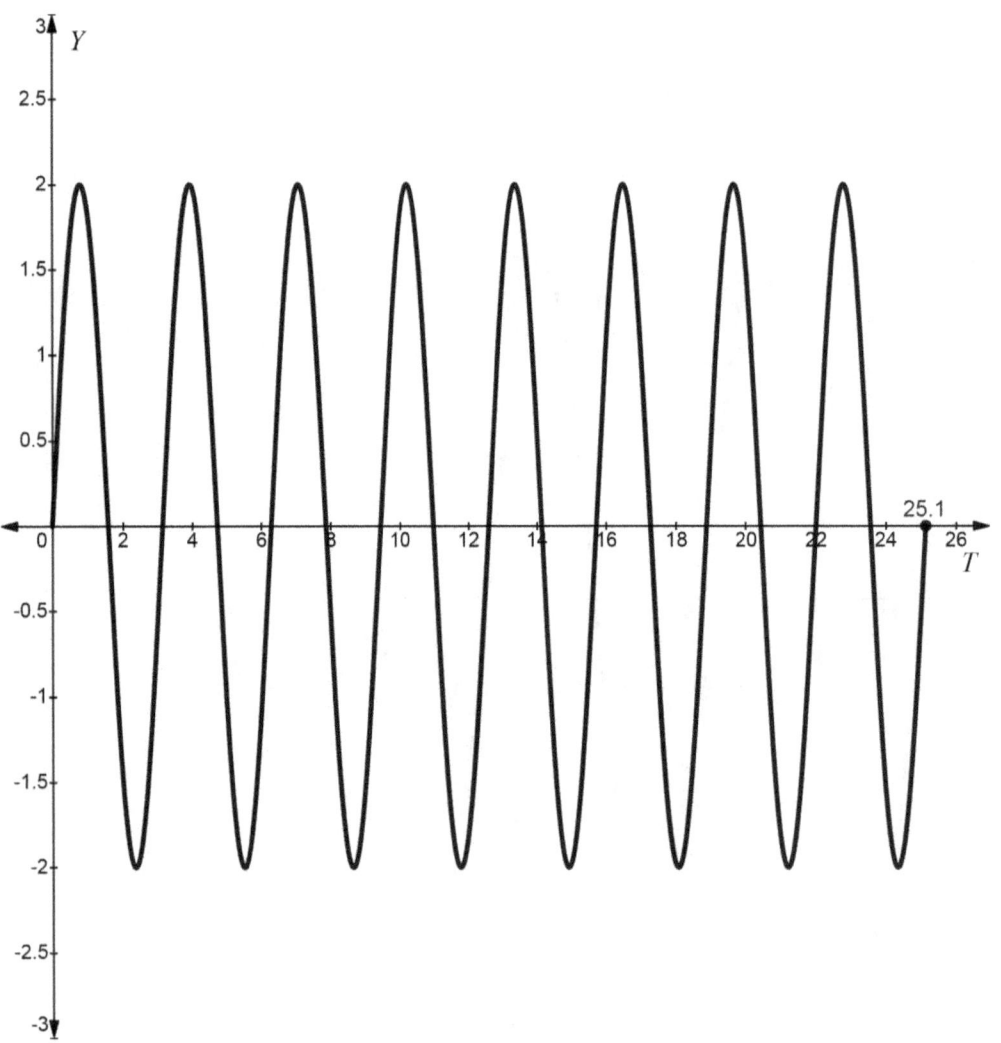

Figure 17 is similar to Figure 16 in that it has the same amplitude of 2 units and the same time frame of 25.1 seconds. But the change from Figure 16's frequency of .159 Hz to Figure 17's frequency

48

of .318 Hz is clearly seen. The wave in Figure 16 completes four cycles in 25.1 seconds. Because its frequency is twice as great, the wave in Figure 17 completes twice as many cycles, eight, in 25.1 seconds.

Equation 29 shows us that the multiplication of frequency, "f", by time (duration), "t", and "2π", equals radians, "B". Substituting Equation 29 into Equations 23 and 24 gives us more control over the Cosine wave equation, Equation 30, and the Sine wave equation, Equation 31, because, after we have made the substitution, we can increase the angular radian measure by increasing the frequency, "f", by increasing the duration, "t", or by both.

Note that the radian measure "2πf t" follows the convention discussed above. A positive value of "2πf t" denotes counterclockwise movement around a circle while a negative value denotes clockwise movement.

There is one more variable we need to add to Equations 30 and 31 that is needed in order to allow us to create waves of any shape. That variable is "phase", which we briefly introduced above. The phase difference of two waves is the amount that one wave is offset against the other wave, measured in radians. Figure 13, Table 1 and Figure 14 above showed us that graphing the y-axis values of a circle of radius 2 units for one complete revolution around the circle plots a Sine wave from 0 to 2π radians whose amplitude is 2.

Suppose we had added the radian value "π/4" to each value along the circle in Figure 13 and proceeded to plot the resulting points as y-axis values until we had completed one 2π revolution around the circle. (0 radians becomes π/4 radians, π/4 radians becomes π/2 radians, and so on.) Figure 18 compares the new graph with Figure 13:

Figure 18

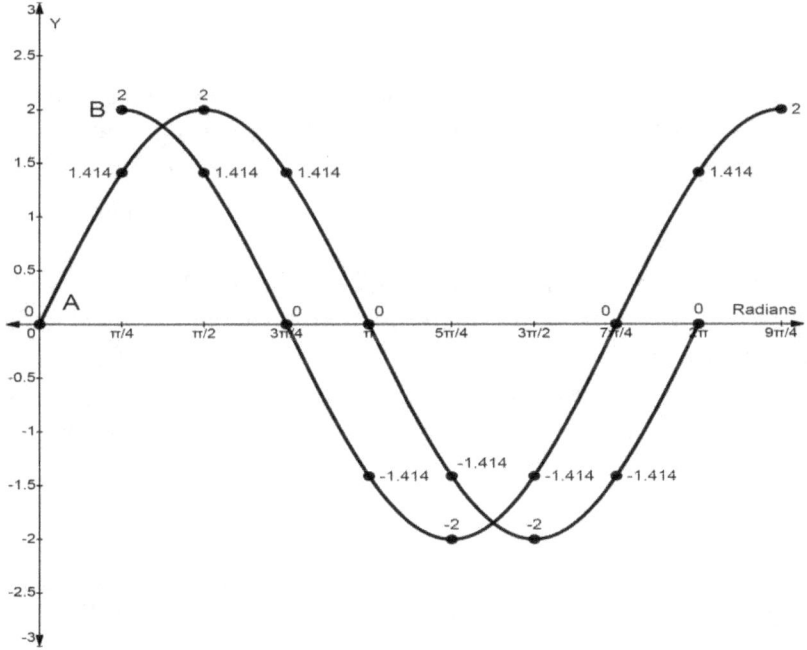

Figure 18 contains two Sine waves. The first Sine wave, wave "A", starts at the origin of the graph (0, 0) and traverses 2π radians until it arrives at point (2π, 0). This is the Sine wave depicted in Figure 14. The second Sine wave, wave "B", starts at point ($\pi/4$, 2) and traverses 2π radians until it arrives at point ($9\pi/4$, 2):

32. $9\pi/4 - \pi/4 = 8\pi/4 = 2\pi$

We can see from Figure 18 that wave B is always "ahead" of wave A by $\pi/4$ radians. For example, wave B starts out at a value of 2 when its radian measure is $\pi/4$ while wave A has a value of 1.414 when its radian measure is $\pi/4$. Wave A does not achieve a value of 2 until its radian measure increases by $\pi/4$, to $\pi/2$. Similarly, wave B has a value of 0 when its radian measure is $3\pi/4$ whereas wave A does not achieve a value of 0 until $\pi/4$ radians later, at π radians. And so on. This $\pi/4$ discrepancy between when the two waves reach the same value is called the "phase difference" between waves A and B and we say the two waves are "out of phase" by $\pi/4$. Alternatively, we can say that wave B "leads" wave A by $\pi/4$ or that wave A "lags" wave B by $\pi/4$. These are all ways of saying the same thing. Notice that the phase difference between waves A and B does not affect their amplitude.

Equations 33 and 34 incorporate the phase of the Cosine and Sine waves:

33. $x = r \cos(2\pi f\, t + \alpha)$

34. $y = r \sin(2\pi f\, t + \alpha)$

where α (alpha) is the positive phase shift the wave has from the origin of a circle of radius "r" (r, 0 radians), or where a circle of radius "r" first intersects the horizontal x-axis as in Figure 13, when measured in a counterclockwise direction.

Following the convention for radian measure discussed above, when the phase shift is measured in a counterclockwise direction from the origin, it has a positive (+) sign in front of it and when it is measured in a clockwise direction from the origin, it has a negative (-) sign in front of it. The phase shift, α, is almost always treated as a constant in the Cosine and Sine wave equations, but there is no reason it cannot vary over time for a complicated wave.

Equation 33 is the horizontal x-axis component of a point whose radian value is "$2\pi f\, t + \alpha$" on a circle of radius "r". Equation 34 is the vertical y-axis component of a point whose radian value is "$2\pi f\, t + \alpha$" on a circle of radius "r". This description assumes that the circle is drawn on the x, y Cartesian Coordinate System as in Figure 13. Since we know that a point on the Cartesian Coordinate System is designated as (x, y) or (x-axis coordinate, y-axis coordinate), it follows from Equations 33 and 34 that:

35. $(x, y) = (r \cos(2\pi f\, t + \alpha),\ r \sin(2\pi f\, t + \alpha))$

Equation 33 tells us that any Cosine wave of amplitude "r", frequency "f", duration "t" and phase "α" can be constructed from a point that starts out at positive radian position "α" on a circle of radius "r" and then moves in a counterclockwise direction around the circle at frequency "f" for duration "t". Equation 34 tells us that any Sine wave of amplitude "r", frequency "f", duration "t" and phase "α" can be constructed from a point that starts out at positive radian position "α" on a circle of radius "r" and then moves in a counterclockwise direction around the circle at frequency "f" for duration "t". Finally, Equation 35 gives us the x-axis and y-axis coordinates of the point at any time "t" that it is moving around the circle. For clockwise movement around a circle, one or both of the terms in Equation 35 ("2πf t", "α") may have a negative sign in front of it.

The takeaway from the above is that waves can be described as points moving around circles. We can describe waves of any shape by specifying the radius of the circle, which becomes the amplitude of the wave, where on the perimeter of the circle a point begins moving around the circle in a clockwise or counterclockwise direction, or the phase of the wave, how "fast" the point is moving around the circle, or the frequency of the wave, and, finally, how long the point is moving around the circle in a clockwise or counterclockwise direction, or the duration of the wave.

In Equation 35 above, the "x" component of a moving point on the circumference of a circle of radius "r" involves the Cosine function while the "y" component involves the Sine function. That is the only thing that is different between the "x" and "y" components. What is the difference between a Sine wave and a Cosine wave? They are out of phase as seen in Figure 19:

Figure 19

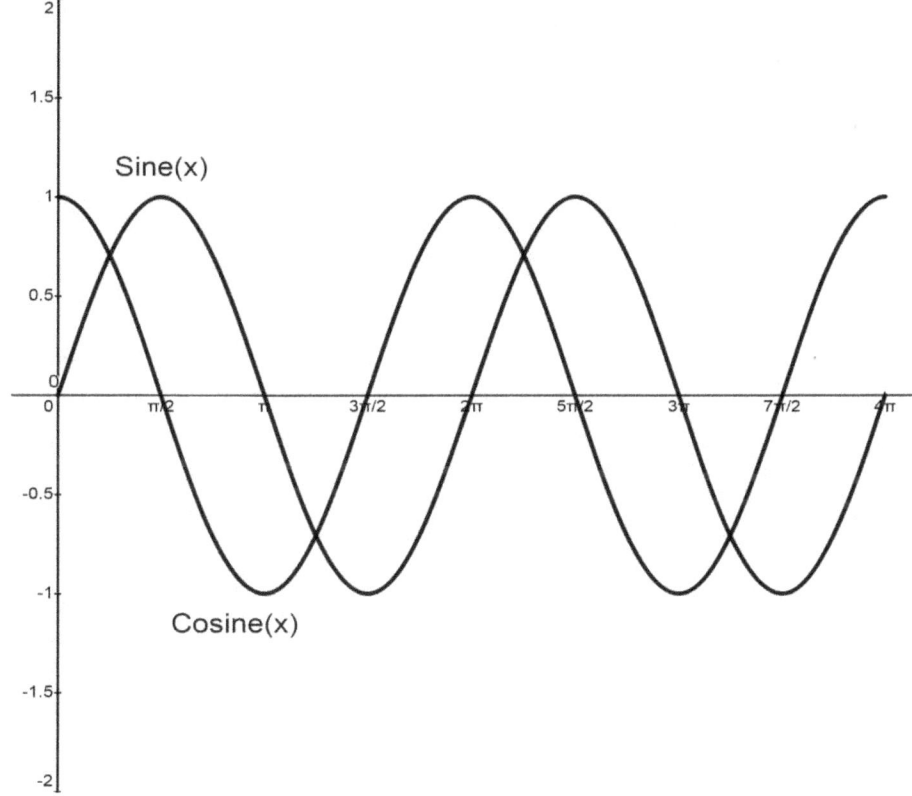

Figure 19 shows a Sine wave and a Cosine wave plotted over a distance of 4π radians, or going twice around the unit circle. Whether or not the Cosine wave leads the Sine wave or the Sine wave leads the Cosine wave is arbitrary and depends on where you begin on the circle. If you start at value "1", then the Cosine wave leads the Sine wave by $\pi/2$ radians because the Cosine wave has the value of "1" at 0 radians while the Sine wave does not reach the peak value of "1" until radian value $\pi/2$.

On the other hand, if you start at value "0", then the Sine wave leads the Cosine wave by $3\pi/2$ radians because the Sine wave has the value of "0" at 0 radians while moving towards its peak while the Cosine wave does not have a value of "0" while moving towards its peak until radian value $3\pi/2$. Note that the waves must be compared while they are moving in the same direction. The Cosine wave actually first reaches the value of "0" at $\pi/2$ radians, but the wave is moving down at that point while the Sine wave had a "0" value at 0 radians while moving up.

Their phase difference is "built into" the Sine and Cosine waves. The "α" (phase) term for the Cosine wave, Equation 33, and the "α" (phase) term for the Sine wave, Equation 34, are both "0" in Figure 19, but there is still a phase difference between the waves. You can correct for the natural phase difference between Cosine waves and Sine waves by adjusting the "α" phase difference of either Equation 33 or Equation 34, but not both. For example, if you let "α" be "$\pi/2$" in the Sine wave equation, Equation 34, and leave "α" at "0" in the Cosine wave equation, Equation 33, the waves graphed in Figure 19 would be the same.

Equation 35 and Figure 19 have two waves of differing phase, a Sine wave and a Cosine wave. Our discussion thus far has focused on "brainwaves". Just what do Sine and Cosine waves have to do with brainwaves? The answer is that brainwaves **are** Sine and Cosine waves. In fact, a branch of mathematics called "Fourier Analysis", named after Joseph Fourier, the man who invented and published it in 1807, states that **any** mathematical curve, representing any event, can be decomposed into a collection of Sine or Cosine waves of different frequencies. In the following, we will use the terms "brainwaves" and "Sine" or "Cosine" waves interchangeably.

CHAPTER 7

EULER'S NUMBER: THE CONSTANT "e"

The Constant e

There are constants that occur again and again in our everyday lives. One constant many people are aware of, and one we have used extensively in this book, is the constant pi, or π. The constant π is the ratio of a circle's circumference to its diameter. This ratio exists for all circles. If the ratio of a shape's circumference to its diameter is not π, then the shape is not a circle. Pi is an irrational number because it cannot be expressed as the ratio of two integers. Although it is often expressed as 3.1416, there is an infinite string of numbers to the right of the decimal point. Some mathematicians spend their careers trying to find the next decimal digits of π.

Another constant is "e", also an irrational number that has an infinitely long decimal component. As we noted above, "e" is called "Euler's Number" after Leonard Euler, the mathematician who discovered it. Euler died in 1783. "e" is also called "Nature's Number" because it ubiquitously shows up time and time again in the physical world. "e" is involved in the shapes of galaxies, sunflowers and sea shells. It is used in the calculation of interest rates. It is used to calculate the half-life of radioactive elements. It plays a role in the growth of bacteria and in temperature-activated chemical reactions. "e" is used in the equations of statistics and quantum theory.

The value of "e" is approximately 2.7183. The calculation of "e" is not as intuitive as the calculation of pi. There are alternate ways to calculate "e". One way is Equation 36:

36. $\quad e = \lim_{n \to \infty} (1 + 1/n)^n$

Equation 36 tells us that "e" is the limit as the positive integer "n" approaches infinity (∞) of the quantity "1+1/n" raised to the "nth" power. The reader may be thinking that "e" must approach the number "1" as "n" approaches infinity because "1+1/n" approaches "1" as "n" approaches "∞". However, the power "n" gets larger and larger as "n" approaches "∞" which offsets the "1+ 1/n" part of the equation getting smaller and smaller as "n" approaches "∞". The result of a declining base (1+1/n) being raised to a larger exponent (n) results in sort of a "standoff" between the competing elements of Equation 36, which can't get any larger than approximately 2.7183 as "n" approaches "∞". Note that the limit is as "n" **approaches** "∞" and the calculation does not include "n" **being equal** to "∞". If it did, the "n" of Equation 36 would become infinite and the calculation of "e" would be undefined mathematically. Table 2 shows some values in the calculation of "e":

Table 2

n	e
1	2.00000
2	2.25000
5	2.48832
10	2.59374
100	2.70481
1000	2.71693
10,000	2.71815
100,000	2.71827

Table 2 shows that the calculation of "e" approaches a limit as "n" approaches infinity. For example, as "n" increases by 99 from 1 to 100, the calculated value of "e" increases significantly from 2.00000 to 2.70481 for an increase of .70481. On the other hand, the bottom of Table 2 shows that as "n" increases by 90,000 (from 10,000 to 100,000), the calculated value of "e" increases from 2.71815 to 2.71827 for an increase of only .00012. As "n" increases substantially, the calculated value of "e" barely increases, indicating that the calculated value of "e" is approaching a limit (which it never reaches).

Table 3 lists some values of "e" raised to various powers:

Table 3

x = power of e	e^x	value
0	e^0	1
1	e^1	e ($\approx$ 2.7183)
2	e^2	7.39
3	e^3	20.09
4	e^4	54.60
5	e^5	148.41
6	e^6	403.43
2π	e^(2π)	535.49
$2\pi i$	e^($2\pi i$)	?

Table 3 raises "e" to increasing exponents. (Note that the term "^" means that the number, term or expression following it is an exponent.) Remembering our Algebra 1 class, we know that raising any number to the "0" power results in a value of "1", so e^0 equals "1". And raising any number to the first power ("1") results in the same value as the number, so e^1 equals "e". Table 3 shows that, as expected, raising "e" to the increasing real (non-imaginary) powers in the left column results in increasing values in the right column. Raising "e" to the 6th power, or e^6, results in value of approximately 403.43. "2π" is just a number (approximately 6.28) and we know from our previous discussion that it is the circumference of a unit circle. Table 3 shows that raising "e" to "2π" results in a value of 535.49.

Unlike the other entries in column 1 of Table 3, the last entry, "$2\pi i$", is not a real number. It is an imaginary number because it includes "i". "i" is defined as the square root of minus 1, or $\sqrt{-1}$. (Note that the term "$\sqrt{}$" means that the square root of the number, term or expression following it shall be taken.) We have used "i" above in defining imaginary time. "$2\pi i$" represents an imaginary unit circle. So what is "e" raised to the imaginary unit circle, or $e^{(2\pi i)}$? Incredibly, the value of "e" raised to the imaginary unit circle is 1! This result is given in Equation 37:

37. $e^{(2\pi i)} = 1$

Equation 37 has been called by mathematicians "the most beautiful equation" and "the most remarkable formula in mathematics". It connects the most ubiquitous constants in nature, "e" and "π", i (the square root of minus 1) and the numbers 1 and 2. It also includes the number "0" because the subtraction of Equation 37's right side from its left side equals "0":

38. $e^{(2\pi i)} - 1 = 0$

Equation 37 repeats every 2π:

37. $e^{(2\pi i)} = 1$

       $[e^{(2\pi i)}]^n = 1^n$ Raise both sides of Equation 37 to integer "n".

39. $e^{(n2\pi i)} = 1$ Algebra exponent rules simplify both sides.

Equation 39 tells us that $e^{(2\pi i)} = 1$, $e^{(4\pi i)} = 1$, $e^{(6\pi i)} = 1$, and so on.

CHAPTER 8

THE IMAGINARY CIRCLE AND EULER'S FORMULA

Equation 37 deals with "e" raised to an imaginary 2π radians ("i2π", which is often expressed as "2πi"). Let's consider the more general relation where the radian value θ (theta) can be any value:

40. $f(\theta) = e^{\wedge}(i\theta)$

where "θ" is the radian value (measured in arc length or time) of any point along the perimeter of the imaginary unit circle, just as it was the radian value of any point along the perimeter of the non-imaginary unit circle discussed above.

In Equation 40, "$f(\theta)$" means "$e^{\wedge}(i\theta)$ is a *function* of θ". Figure 20 is the graph of Equation 40, or $e^{\wedge}(i\theta)$:

Figure 20

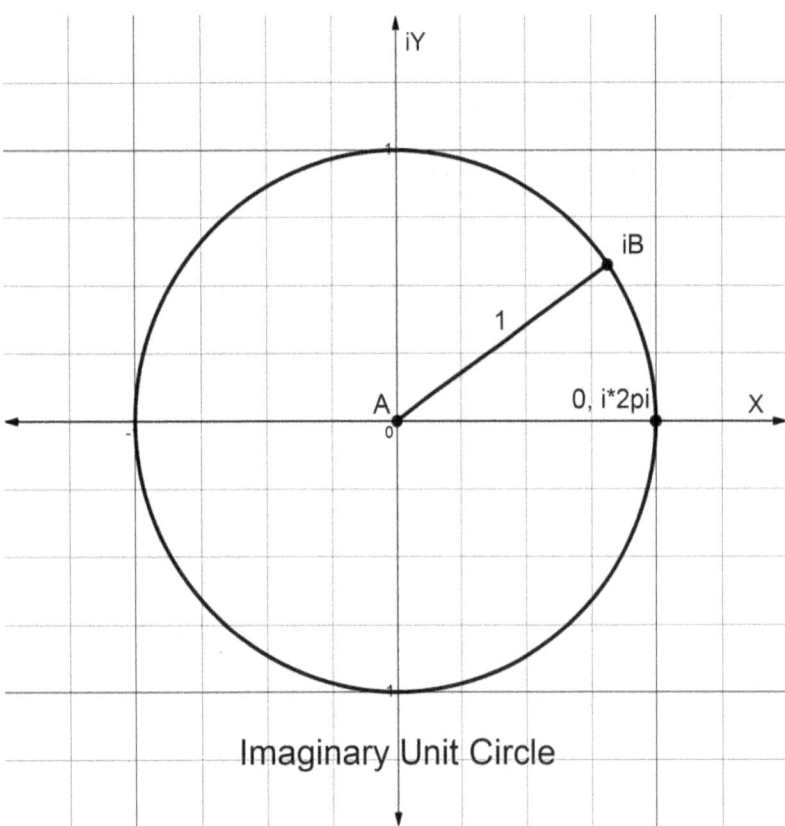

The graph of Equation 40 is the imaginary unit circle! It differs from the (non-imaginary) unit circle we discussed previously and seen in Figure 11 in only two respects. First of all, while the horizontal x-axis is the same as in Figure 11, the vertical y-axis is now imaginary with the "Y" of Figure 11 being replaced by the "iY" of Figure 20. Secondly, while the radian arc lengths in Figure 11 are in real (non-imaginary) numbers, the arc lengths in Figure 20 are now imaginary as denoted by the "i" contained in them. This can be seen where the real arc length "B" of Figure 11 has become "iB" in Figure 20.

The imaginary unit circle starts out at coordinate point (1, 0) where it intersects the circle along the horizontal x-axis. This point is called the "origin" of the circle. The arc length of a point sitting on the origin is 0. As the point traverses the imaginary unit circle counterclockwise, it completes one $i2\pi$ revolution back to the origin. If it completes another clockwise revolution around the circle it will have traversed a total distance of $i4\pi$. Yet another revolution will be a total distance of $i6\pi$, and so on as seen in Equation 39.

The radius of the circle in Figure 20 is "1". That is why Figure 20 is called the imaginary "unit" circle, the unit being "1". But, just as in the case of the real unit circle discussed above, there is no reason an imaginary circle cannot have a radius bigger or smaller than "1". It just won't be an imaginary *unit* circle. Figure 13 showed a real circle of radius 2 along with various radian arc lengths around it. Figure 21 shows a similar figure for an imaginary circle of radius 2:

Figure 21

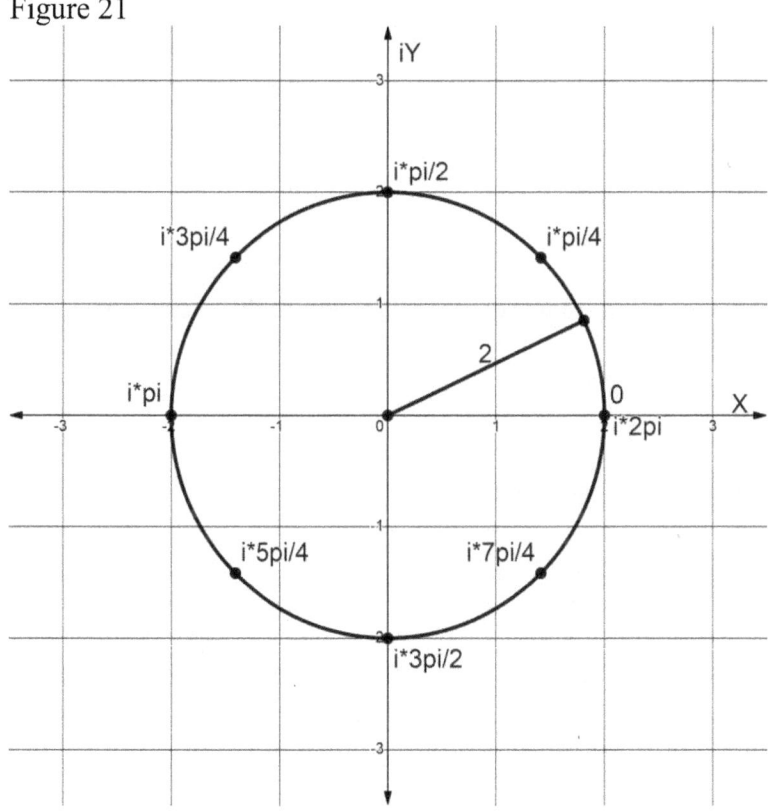

In Figure 21, "π" is denoted as "pi".

The equation for an imaginary circle with radius "r" is given by Equation 41:

41. $f(\theta) = r\, e^{(i\theta)}$

where "r" is the radius of the imaginary circle, and "θ" is the radian value from where the imaginary circle intersects the horizontal x-axis to a point on the perimeter of the imaginary circle. Equation 41 is just Equation 40 with an "r" in front of it. Note that, when "r" is "1", Equation 41 becomes Equation 40. Note also that "r" can be bigger *or smaller* than "1".

Regardless of the length of its radius, a point on the perimeter of a real circle will always traverse 2π real (non-imaginary) radians on a round trip from its starting point completely around the circle back to the starting point whereas a point on the perimeter of an imaginary circle will always traverse 2π **imaginary** radians (2πi) on a similar round trip.

In an imaginary circle derived from Equation 41, there will always be two real components and two imaginary components. The real components are the radius and the horizontal x-axis coordinate of any point on the perimeter of the circle. The imaginary components are the radian values, denoted in Figure 21 with an imaginary "i" contained in them (e.g. - iπ/4, i3π/2, etc.) and the vertical y-axis coordinate of any point on the perimeter of the circle. (Note the "i" in the "iY" vertical y-axis label.)

We recall that the horizontal x-axis of a point on a real circle is given by Equation 23:

23. $x = r\cos(B)$

where:

"x" is the horizontal-axis component of the point whose radian value is "B".
"r" is the radius of the circle.

We further recall that the vertical y-axis of a point on a real circle is given by Equation 24:

24. $y = r\sin(B)$

where:

"y" is the vertical-axis component of the point whose radian value is "B".
"r" is the radius of the circle.

Since we are now dealing with imaginary circles, the reader may be wondering what the x-axis and iy-axis coordinates are of a point on an imaginary circle. The answer is called "Euler's Formula", which was discovered by mathematician Leonhard Euler around 1740. It is considered to be one of the most amazing mathematical results ever discovered. Euler's Formula, modified to include the radius of a circle, "r", is given by Equation 42:

42. $r e^{i\theta} = r \cos(\theta) + i r \sin(\theta)$

where:

"r" is the radius of an imaginary circle.
"θ" is the radian value from where the imaginary circle intersects the horizontal x-axis to a point resting on the perimeter of the imaginary circle.

The right-hand side of Equation 42 is shorthand for the x-axis and y-axis coordinates of an imaginary number, so we can say:

43. x-axis coordinate of $r e^{i\theta} = r \cos(\theta)$

   y-axis coordinate of $r e^{i\theta} = i r \sin(\theta)$

or:

44. $r e^{i\theta} = (r \cos(\theta), i r \sin(\theta))$

Equation 44 tells us that "$r e^{i\theta}$" is a point on the perimeter of an imaginary circle of radius "r", that the "point" is the point that has radian value "iθ", that the x-axis coordinate of the point is "$r \cos(\theta)$" and that the imaginary y-axis ("iy") coordinate of the point is "$i r \sin(\theta)$".

We would provide a proof of Equation 44, if there was one! Equation 44 is an example of a "Voila!" result that comes along once in a great while. Someone thinks of something without formal proof and "Voila!", it works. (Another example of this can be found in quantum theory.) Equation 44 provides the coordinates of a stationary point on the perimeter of an imaginary circle of radius "r". What if the point were moving around the imaginary circle?

In Equations 33 and 34, we showed that we could substitute "$2\pi f t + \alpha$" for a stationary radian value of a point on a real circle in order to describe the point as it became non-stationary and moved around the circle starting at phase "α" radians for duration "t" seconds at a frequency of "f" Hz, or cycles per second. The quantity "$2\pi f t + \alpha$" is in radians (2π radians/cycle * f cycles/second * t seconds + α radians) and substitutes the stationary radian value for a moving one. Since the "θ" in "$r e^{i\theta}$" is also in radians, we can substitute "$2\pi f t + \alpha$" for it in Equation 44:

44. $r e^{i\theta} = (r \cos(\theta), i r \sin(\theta))$

45. $r e^{i(2\pi * f * t + \alpha)} = [r \cos(2\pi f t + \alpha), i r \sin(2\pi f t + \alpha)]$

Equation 45 tells us that a point moving around an imaginary circle creates two waves: a real Cosine wave of radius "r" which starts out at phase "α" and moves in a counterclockwise direction at a frequency of "f" Hz for a duration of "t" seconds and an imaginary Sine wave of radius "r" which starts out at phase "α" and moves in a counterclockwise direction at a frequency of "f" Hz for a duration of "t" seconds. If we wanted the waves to move in a clockwise direction from phase "α", then the first term in the Cosine and Sine functions, "$2\pi f t$", would have a minus sign in front of it, or "$-2\pi f t$".

A comparison of the imaginary circle wave equation, Equation 45, and the real circle wave equation, Equation 35:

35. $(x, y) = (r \cos(2\pi f\, t + \alpha), r \sin(2\pi f\, t + \alpha))$

shows us that the real and imaginary circle wave equations differ only in the "i" before the Sine wave function. In other words, the real and imaginary circle Cosine wave equations are the same and produce real waves:

$r \cos(2\pi f\, t + \alpha)$ real and imaginary circle Cosine wave equations

while the real circle Sine wave equation produces a real Sine wave:

$r \sin(2\pi f\, t + \alpha)$ real circle Sine wave equation

and the imaginary circle Sine wave equation produce an imaginary Sine wave:

$i\, r \sin(2\pi f\, t + \alpha)$ imaginary circle Sine wave equation.

So making a radian value imaginary and raising it to a power of "e", or "e^(iθ)", creates an imaginary Sine wave and a real Cosine wave.

CHAPTER 9

HYPERBOLIC SPACETIME

Euler's Formula Equations 44 and 45 tell us that raising "e" to an imaginary exponent results in the creation of two waves, a real wave and an imaginary wave of the same amplitude (r), frequency (f), duration (t) and phase (α) and that the imaginary wave and real waves are out of phase with one another. (We saw above that Cosine waves and Sine waves are out of phase.) The real wave exists in the present and the imaginary wave occurs in the future. "Real" *means* "present" and "imaginary" *means* "future". Recall from above that "imaginary time" and the "future" are synonymous. What does all of this tell us?

It *appears* to tell us that, if we could travel to the future *and* find an imaginary circle to "hitch a ride on" once we got there, we could create imaginary brainwaves (Sine waves) describing any event we chose to create and then experience that event in the future during an "inside" event OBE.

The first part is easy and then we run into problems. We have seen that we can travel to the future by shutting off our sensors. Let's say we do so and then we are able to find an imaginary circle. Once we find the imaginary circle, we can move around its perimeter, creating an imaginary Sine wave as we do so. But we also simultaneously and automatically create a real cosine wave of the same frequency, "f", duration, "t", and starting point, "α", as the imaginary Sine wave we created but is out of phase with the imaginary Sine wave as we have seen in Figure 19. We don't need or want the Cosine wave. The Cosine wave occurs in the present. But an OBE is an inside event that occurs entirely in the future. In addition, because the real Cosine waves generated by us while we are on the imaginary circle are out of phase with the imaginary Sine waves, they will not convey the exact same event that our imaginary waves created.

This means that the imaginary Euler Formula circle does not quite model, or adequately explain, OBEs. Close, but not quite. It is not possible to move around the perimeter of the imaginary Euler Formula circle to simultaneously create imaginary and real waves that are in phase with one another. The real wave is the sequence of horizontal x-axis components of the points on the perimeter of the imaginary circle that we traverse while the imaginary wave is the sequence of the vertical y-axis components that we simultaneously traverse. In order to move around an imaginary circle and simultaneously create in-phase imaginary and real waves, the x-axis and y-axis coordinates of every point we traverse would have to be the same. A perimeter point's "x" coordinate would always have to equal its "y" coordinate. But if this were the case, we would not be on the perimeter of a circle. This can be confirmed by looking at Figure 20, repeated here:

Figure 20

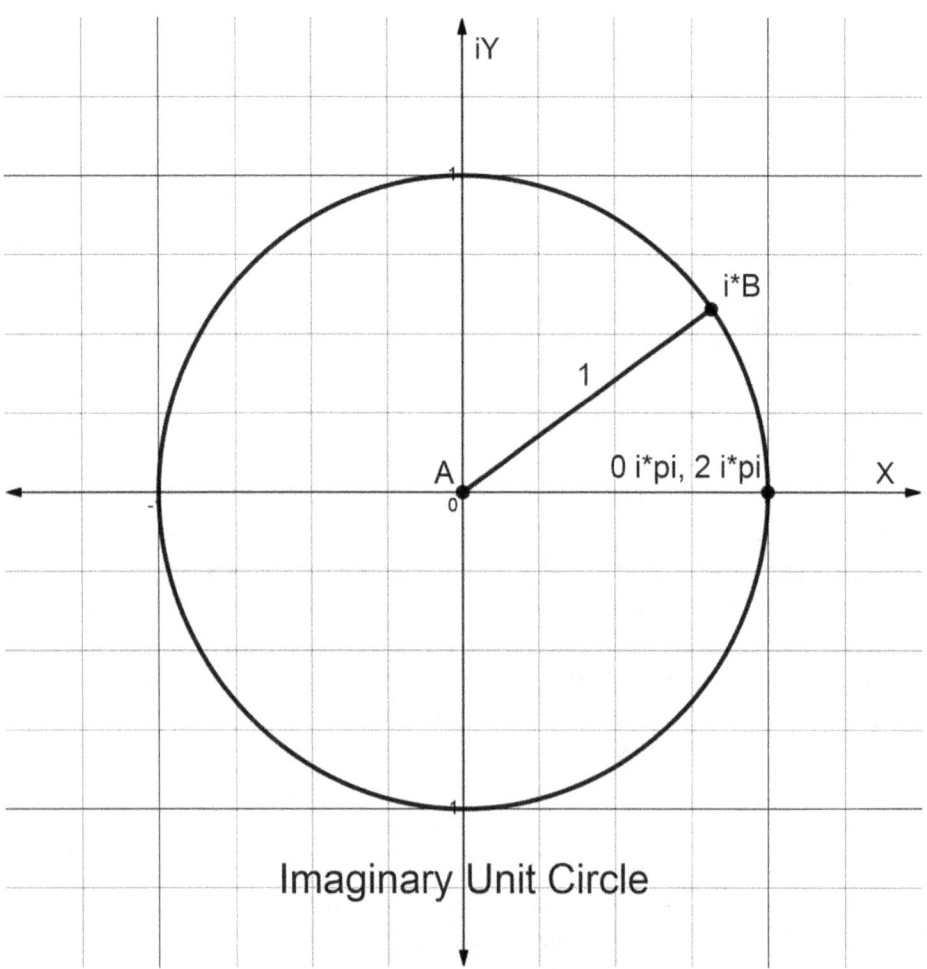

Imaginary Unit Circle

For example, as "X" goes from "0" to "1" in Figure 20, "Y" goes in the opposite direction from "1" to "0". The "x" coordinate and "y" coordinate of an imaginary circle (or any circle) are only equal to each other two times, once in the first quadrant when $x = y = (\sqrt{2})/2$ and once in the third quadrant when $x = y = -(\sqrt{2})/2$.

Furthermore, and most importantly, at the root of the problem is the fact that we will not find a Euler's Formula imaginary circle waiting for us after we have turned off our sensors and traveled to the future. Why? Because the "shape" of space and time, or spacetime, is not governed by Euclidian Geometry, which Euler's Formula is part of. Euclidean Geometry assumes both space and time are "orthogonal" or are at right angles to each other. But Einstein's General Theory of Relativity shows that time is not orthogonal to space. The shape of the universe is curved. Time bends space back on itself.

When we are in the present, we exist in space. We can move through space, but we can't move through time. Time passes, but we can't move through it like we can move through space. Euler's

Formula describes, or approximately describes, this Euclidian situation. But when we have an OBE, we exist in imaginary time after we have turned off our sensors and left "space". The imaginary time we have entered is not orthogonal to the space we have left by turning off our sensors and Euler's formula does not apply to it.

Accordingly, we need to find an equation that describes the actual shape of spacetime if we expect to encounter and manipulate it when we have an OBE. Albert Einstein's theories of Special and General Relativity deal with, among other things, the shape of spacetime. The geometry of Special and General Relativity is hyperbolic, or based on the hyperbola. (See, for example, "The Geometry of Special Relativity", 2nd edition, by Tevian Dray, Chapman and Hall/CRC, 6/10/2021 and "Visual Complex Analysis", by Tristan Needham, Clarendon Press, Oxford, 1997.) One of Einstein's former professors, Hermann Minkowski, is credited with the formalization of spacetime around the turn of the 20th century. In "Minkowski spacetime", time and space are not separated entities but are intermingled in a four-dimensional spacetime. The mathematical basis of Minkowski spacetime is also based on the hyperbola.

The takeaway from the above is that spacetime is not (completely) described by Euclidean geometry or the Euler Formula imaginary circle. It is hyperbolic, or described by hyperbolas. In mathematics, hyperbolic functions are similar to ordinary trigonometric functions, but defined using the hyperbola rather than the circle. Just as the points [cos(x), sin(x)] involve a circle with a unit radius, the points [cosh(x), sinh(x)] involve a unit hyperbola. "Cosh(x)" stands for "Hyperbolic Cosine" and "sinh(x)" stands for "Hyperbolic Sine". Figure 22 illustrates a unit hyperbola:

Figure 22

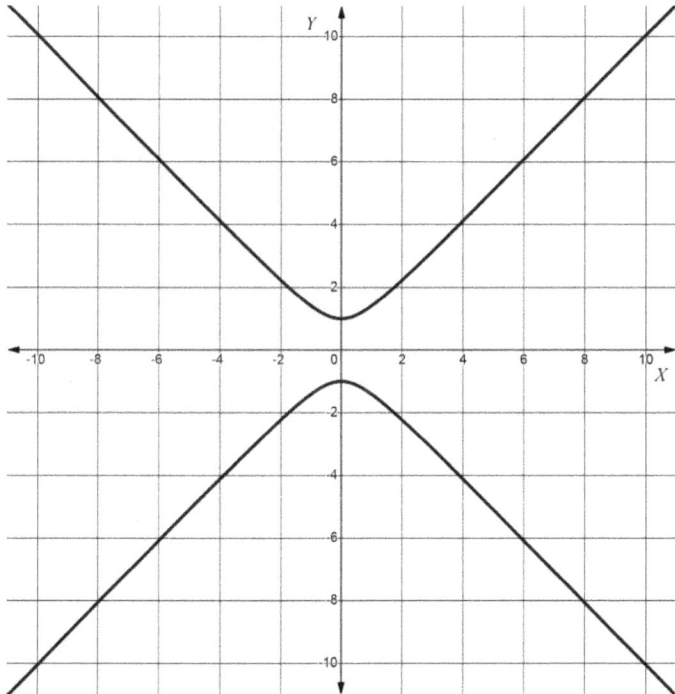

The hyperbola in Figure 22 is a "unit" hyperbola because the low point of the top half of the hyperbola is "1".

The general equation of a hyperbola is:

46. $(y^2)/(a^2) - (x^2)/(b^2) = 1$

where a and b are constants.

When a = b, the equation of a hyperbola:

47. $y^2 - x^2 = a^2$

is very similar to the equation of a circle:

48. $y^2 + x^2 = r^2$

where r^2 is a constant. (r is the radius of the circle.)

Equations 47 and 48 only differ by the sign between the "x^2" and "y^2" terms. Note that, when a = r = 1, Equations 46 and 47 are the unit hyperbola and the unit circle, respectively.

What makes Equation 48, the circle, Euclidian but the very similar Equation 47 hyperbolic? It has to be the difference in the signs of x^2 and y^2 because that is the only difference in the equations. The equation of the hyperbola has different signs for x^2 and y^2 (positive for y^2, negative for x^2) while the circle has positive signs for each of its variables. Any term in a hyperbola with a negative sign in front of it is not orthogonal (at right angles) to the terms with positive signs in front of them.

We have seen that we exist in space and time. We can only *experience* time (clock time) in the present as it flows through us. We have to move to the future by shutting off our sensors in order to experience and *move through* imaginary time, or the future (iT). Space represents the present and is filled with events consisting of brainwaves which are brought to our consciousnesses by our sensors. Imaginary time represents the future and we can shut down our sensors and move to the future in order to have an OBE.

The equation of a hyperbola can have more than two variables. The equation below is a hyperbolic equation that has three variables:

49. $S^2 - V^2 - T^2 = 0$

rearranging terms:

50. $V^2 + T^2 = S^2$

In Equation 49, S represents the magnitude of our sensor input at any given point in time, V represents our visualization at any given point in time and T represents time. T will be either clock time, the present, or imaginary time, the future, depending on the values of the other two variables, S and V. We recall that, if our sensors are turned off, we are able to visualize the future at the "expense" of experiencing the present. If our sensors are on, we experience the present but we can't visualize the future.

The reader may be wondering why V^2 and T^2 have minus signs (-) in front of them in Equation 49. We have seen that the magnitude of our sensor input to our consciousnesses, S, moves in the opposite direction of our ability to visualize the future. Because S has a positive sign in Equation 49, we model this by giving V a minus sign. Also, the magnitude of our sensor input, S, represents space because, as we have seen, our sensors bring space to our consciousnesses. Our sensors, in combination with our brains, most importantly our parietal lobes, literally create the space that we experience. We have seen that T, time, is not orthogonal (not at right angles) to space. This is represented in Equation 49 by also giving T the opposite sign (-) that S has (+). (Note that the + sign of S is not included in Equation 49. A variable that has no sign in front of it is considered to be positive.)

Equation 50 has the same form as Equation 48, the equation for a circle:

48. $y^2 + x^2 = r^2$

where y = V, x = T and r = S

However, the "r" in Equation 48 is a constant equal to the radius of the circle described by the equation, while the "S" in Equation 50 is a variable representing the magnitude of our sensor input at any given point in time. This means that every value that the variable S takes represents a separate circle. As the values of S decline from some higher positive number towards a lower positive number, Equation 50 forms a hyperbolic funnel, or cone. This result is seen in Figures 23 and 24 below, which represent the funnel from different perspectives so the reader can get a good idea of what it looks like:

Figure 23

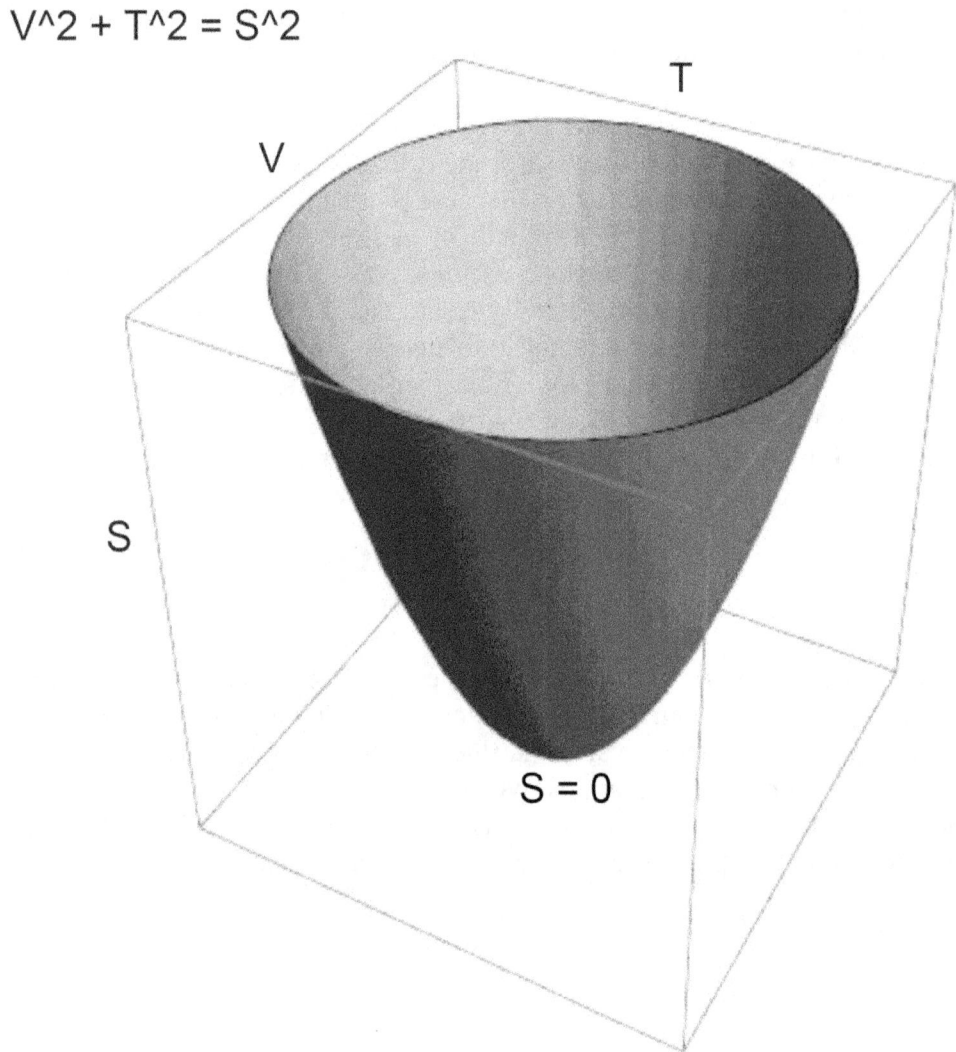

Figure 24

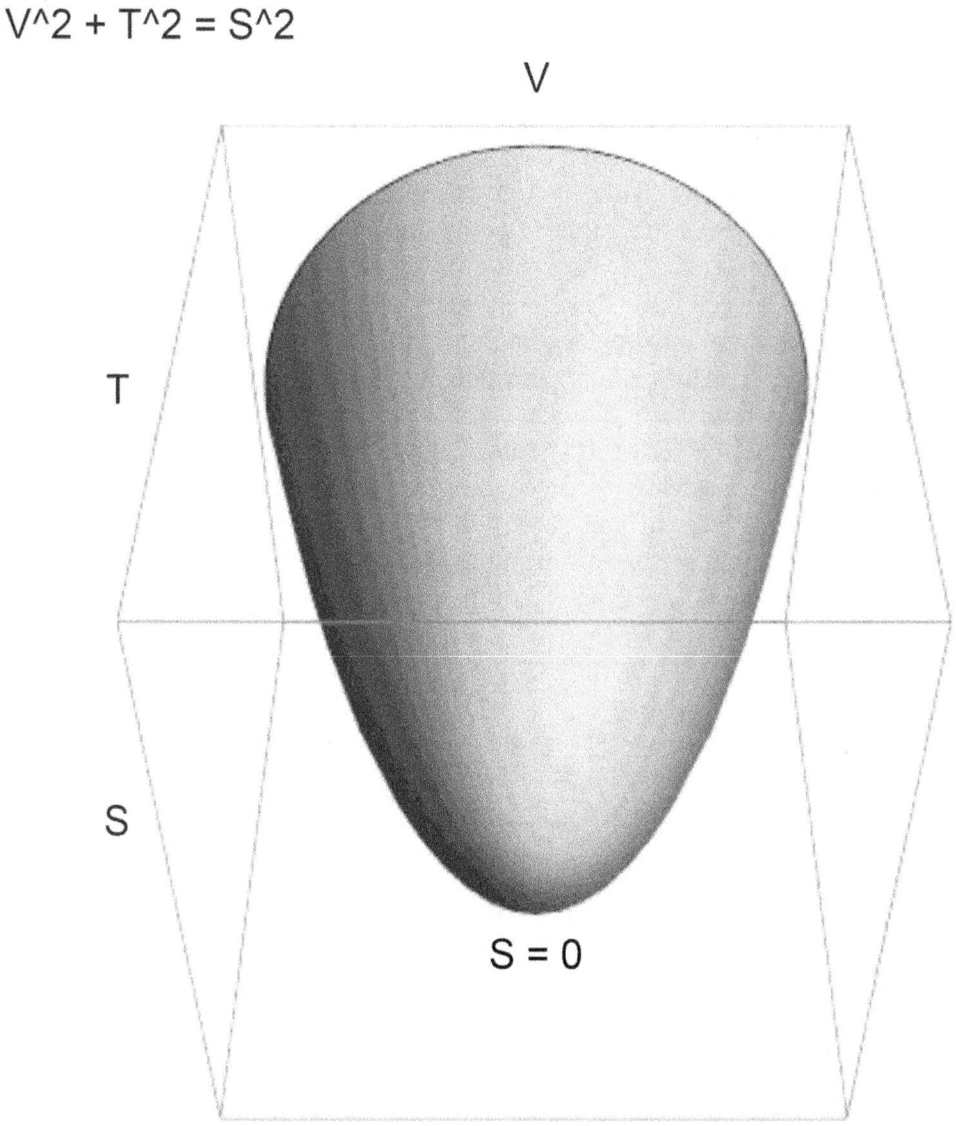

$$V^2 + T^2 = S^2$$

V

T

S

S = 0

Figure 23 shows Equation 50 from the top to the bottom of the object where the sensor input value of "0" appears just below the bottom of the object while Figure 24 shows Equation 50 from the back to the front of the object where the sensor input value of "0" appears just in front of the object. Figures 23 and 24 represent the shape of a hyperbolic funnel composed of individual circles that get smaller and smaller as "S" *approaches* "0". **This is the shape of the funnel that I go down to turn off my sensors and enter imaginary time, or the future.**

Imagine visualizing a funnel shaped like the one in Figure 24. When you first visualize it, your sensors will be active so you will be at its back side where your sensor input has some relatively high positive value. (Note the S-axis in Figure 24. If it appears to you from looking at Figure 24 that the S-axis is vertical instead of horizontal, you are being affected by an optical illusion. Blink your eyes and the S-axis should change from vertical to horizontal!)

After you enter the funnel from its back side, you visualize moving through it. As you do, the tunnel will become narrower and narrower as the value of your sensor input, S, becomes less and less. In other words, you are turning off your sensors as you move down the funnel! You eventually get to the very tip of the funnel where S has a very low value. You keep going and you burst out of the front of the funnel. At his point your sensors will have been completely turned off (S = 0). Note that when S = 0, you are just outside the funnel. You have burst through the front of the tunnel to where your sensors have been shut down.

Why are you outside of the funnel when S = 0 and not just on the tip of it? As we saw above in Equation 50, the funnel is composed of circles with descending radiuses S. When S = 0, there is no circle because a circle must have at least a small non-zero radius. Since there is no circle when S = 0, you can't be part of the funnel. Instead, you are an infinitesimally small distance just outside the funnel. What can we say about your visualization, V, and time, T, at this point? Let's return to Equation 50:

50. $V^2 + T^2 = S^2$

When S = 0, Equation 50 becomes:

$$V^2 + T^2 = 0$$

subtracting T^2 from both sides of the equation

$$V^2 = -T^2$$

taking the square toot of both sides:

$$\sqrt{(V^2)} = \sqrt{(-T^2)}$$

from the rules of Algebra, $(-T^2) = (-1)(T^2)$, so:

$$\sqrt{(V^2)} = \sqrt{[(-1)(T^2)]}$$

from the Algebraic rules of radicals, $\sqrt{[(-1)(T^2)]} = \sqrt{(-1)}\sqrt{(T^2)}$, so:

$$\sqrt{(V^2)} = \sqrt{(-1)}\sqrt{(T^2)}$$

taking the square root of each side of the equation:

$$51. \quad V = iT$$

because the square root of a squared number is the original number and because the square root of -1, or √-1, is the imaginary number "i".

There you have it. **When we visualize moving down and bursting through a hyperbolic funnel such as the one illustrated in Figures 23 and 24, we turn our sensors off and enter imaginary hyperbolic time, or the future.**

Note that $V = iT$ only after our sensors have been completely turned off by traversing to the end of the imaginary funnel and then bursting through it to where $S = 0$. Anytime before that, the radius of the circle, S, will have a positive value greater than T and $V^2 = S^2 - T^2$ will be positive and non-imaginary and therefore so will the value of V.

Equation 51 illustrates the case when our sensors have been turned off. What about the case when our sensors are on in the present and our visualization is off? In that case, the V in Equation 50 will be "0", so:

$$50. \quad V^2 + T^2 = S^2$$

Let $V = 0$:

$$T^2 = S^2$$

Taking the square root of both sides:

$$\sqrt{T^2} = \sqrt{S^2}$$

Simplifying:

$$52. \quad S = T$$

Equation 52 tells us that, when our sensors are on, they "bring" time to us and we experience it in the present as it flows through us. We have called this time "clock time". There is nothing imaginary about it and we can't therefore consciously move through it. When we are in the present, we *experience* time because we are "hooked up" to the outside world via our sensors. Suppose we had no sensors. No eyes, no nose, no ears, no tongue, no skin. Could we experience clock time? No, we couldn't. We would have no connection to the outside world of the present where clock time exists. Think of when you are sleeping and your sensors are relatively inactive. We don't experience clock time then. We go to sleep and wake up several hours later. We have no awareness of how much time passed while we were sleeping. If you are like me, you have to actually check a clock on your bed stand to find out how long you have been asleep.

Equation 51 tells us that, when we are able to turn off our sensors, we are in the future and we can move through the future, or imaginary time, by varying what we are visualizing, V. The equation tells us that, when we turn off our sensors and visualize an event, we place it in the future. The events we create through our visualizations are the "footprints" of our tracks through imaginary time. Whereas we experience time when it is clock time in the present, we move through time when it is imaginary time, or the future. Equation 52 tells us that when our sensors contain us in our bodies, we experience the clock time that our sensors bring to us. But Equation 51 tells us that when we "leave" our bodies by shutting down our sensors, time becomes imaginary as we move into the future.

We experience time in both the present and the future. As we have discussed, present time is called "clock time" while "imaginary time" is the future. We also visualize in both the present and the future as we can see from Equations 50 and 51 and Figure 24. We can visualize going down a funnel in the present to shut off our sensors. After our sensors have been shut off by bursting out of the end of the funnel, we visualize imaginary time, $V = iT$, or the future. Unlike experiencing time and visualizing, we only experience our sensors in the present as they convert events to the brainwaves that they represent. Our sensors *are* the present.

When a jet airplane travels faster than the sound waves it creates, approximately 770 miles per hour, it is said to have *broken through the sound barrier*. When it does this, it bursts out of a hyperbolic funnel or cone as seen in Figure 25:

Figure 25

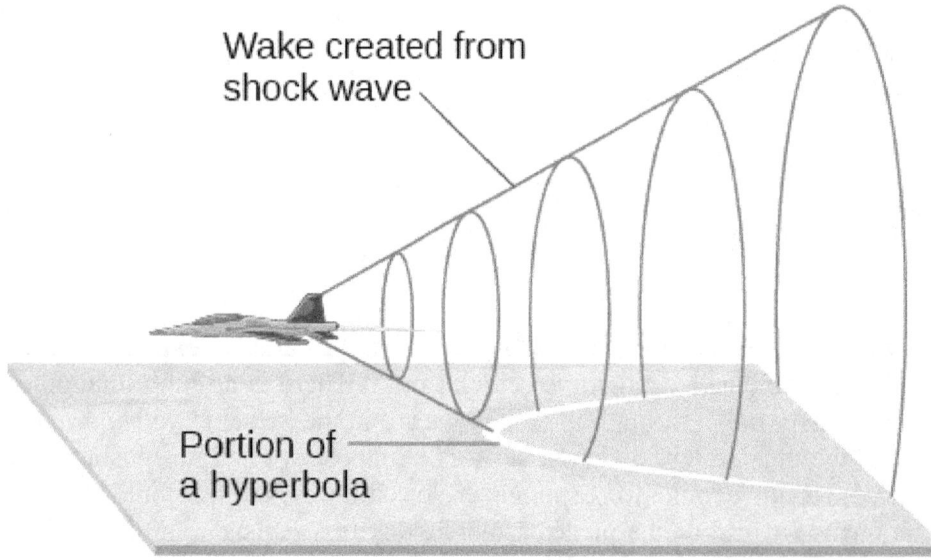

The similarities between Figures 24 and 25 and their descriptions are striking. We have discussed how we enter imaginary time, iT, by bursting out of a hyperbolic funnel. In essence, when we have burst through the hyperbolic funnel, or cone, described in Equation 50, shutting off our sensors in the process, we have *broken through the time barrier and moved from the clock time of the present to imaginary time, or the future.*

Since hyperbolic Equation 50 and its derivatives Equation 51 and Equation 52 accurately "explain" the relationship between our visualization, V, our sensors, S, clock time, T, and imaginary time, iT, the conclusion that presents itself is that we do not exist in Euclidian spacetime. We exist in hyperbolic spacetime!

CHAPTER 10

CREATING WAVES IN HYPERBOLIC TIME

Our mission is not to provide a treatise on hyperbolic geometry. We have spent a scant several paragraphs on it. Our mission is trying to find a mathematical explanation for our ability to create Sine waves in the future. We know from Equation 51 above that we move into the hyperbolic future after we have shut down our sensors. But when we get there, is there a way for us to create Sine waves? In other words, we need to see if there is a hyperbolic equation for an imaginary Sine wave, or brainwave. There is! Equation 53 describes the equation for the imaginary hyperbolic Sine wave:

53. $\sinh(i\theta) = .5[\ e^{\wedge}(i\theta) - e^{\wedge}(-i\theta)\ ] = i\sin(\theta)$

The reader may have been wondering why we went to the trouble of introducing the Euler Formula imaginary circle when the Euler Formula is inadequate to explain OBEs. Equation 53 above shows the reason why. Equation 53 contains two Euler Formula imaginary circles! In effect, Equation 53 corrects the Euclidean space Euler's Formula and turns it into a formula that fits hyperbolic space. And it uses Euler's own Formula to make the correction.

Equation 53 is remarkable. It says that, in hyperbolic spacetime, the spacetime we exist in, the Sine of an imaginary radian value, "θ", is equal to "$i\sin(\theta)$", just as it does in the Euclidian space of Euler's Formula. Furthermore, the equation for the imaginary hyperbolic Sine wave, Equation 53, actually *relies* on the Euclidian space Euler's Formula "$e^{\wedge}(i\theta)$" as we have seen. But the problematic real cosine wave of Euler's Formula has disappeared from $\sinh(i\theta)$. Where did it go?

Before we answer this, we need to establish relationships, called "identities" between $\cos(\theta)$ and $\cos(-\theta)$ and between $\sin(\theta)$ and $\sin(-\theta)$. We choose to do this graphically. Figure 26 contains the graphs of $\cos(x)$ and $\cos(-x)$ ("x" = "θ"):

Figure 26

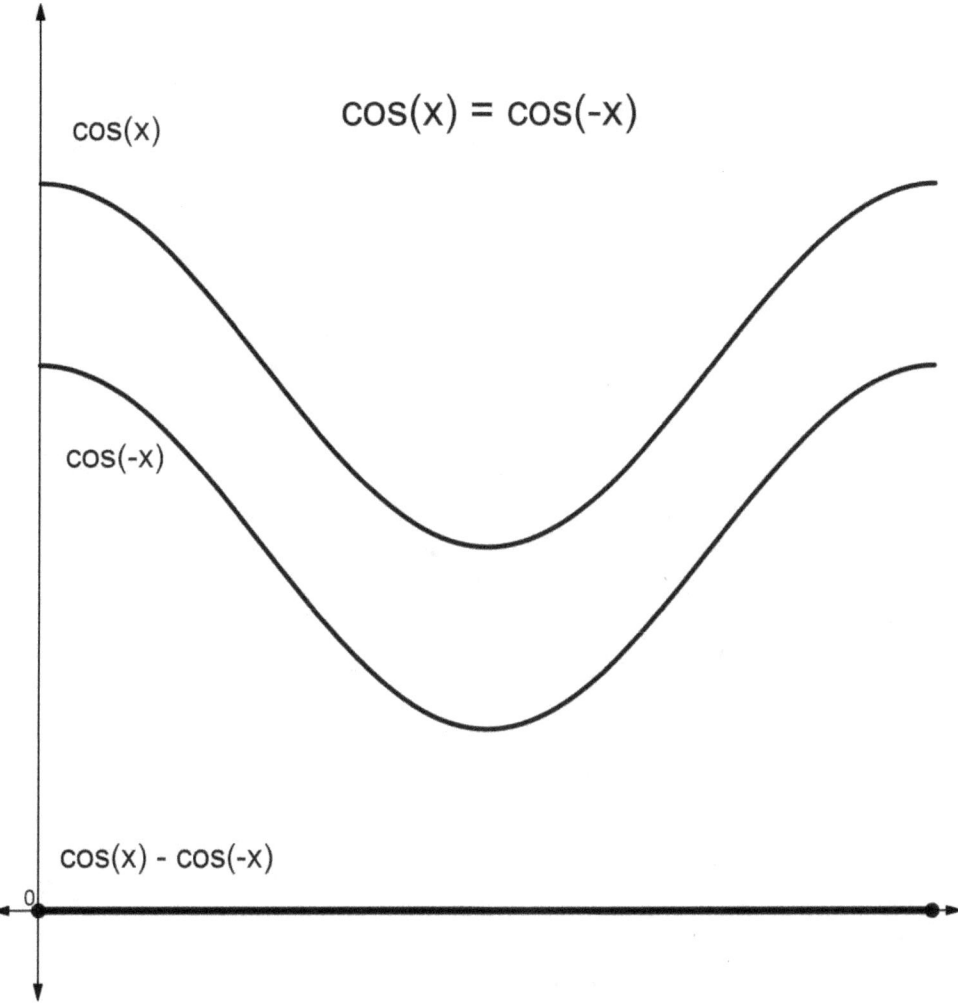

Figure 26 shows the graphs of cos(x) and cos(-x) over the range 0 to 2π. They repeat after that. The graphs have the same shape. The horizontal axis contains the subtraction of the two graphs, cos(x) - cos(-x). Since the vertical component of a line on the horizontal axis is "0", cos(x) must equal cos(-x). **Cos(θ) = cos(-θ)** is a common identity used in trigonometric calculations.

Figure 27 contains the graphs of -sin(x) and sin(-x) ("x" = "θ"):

Figure 27

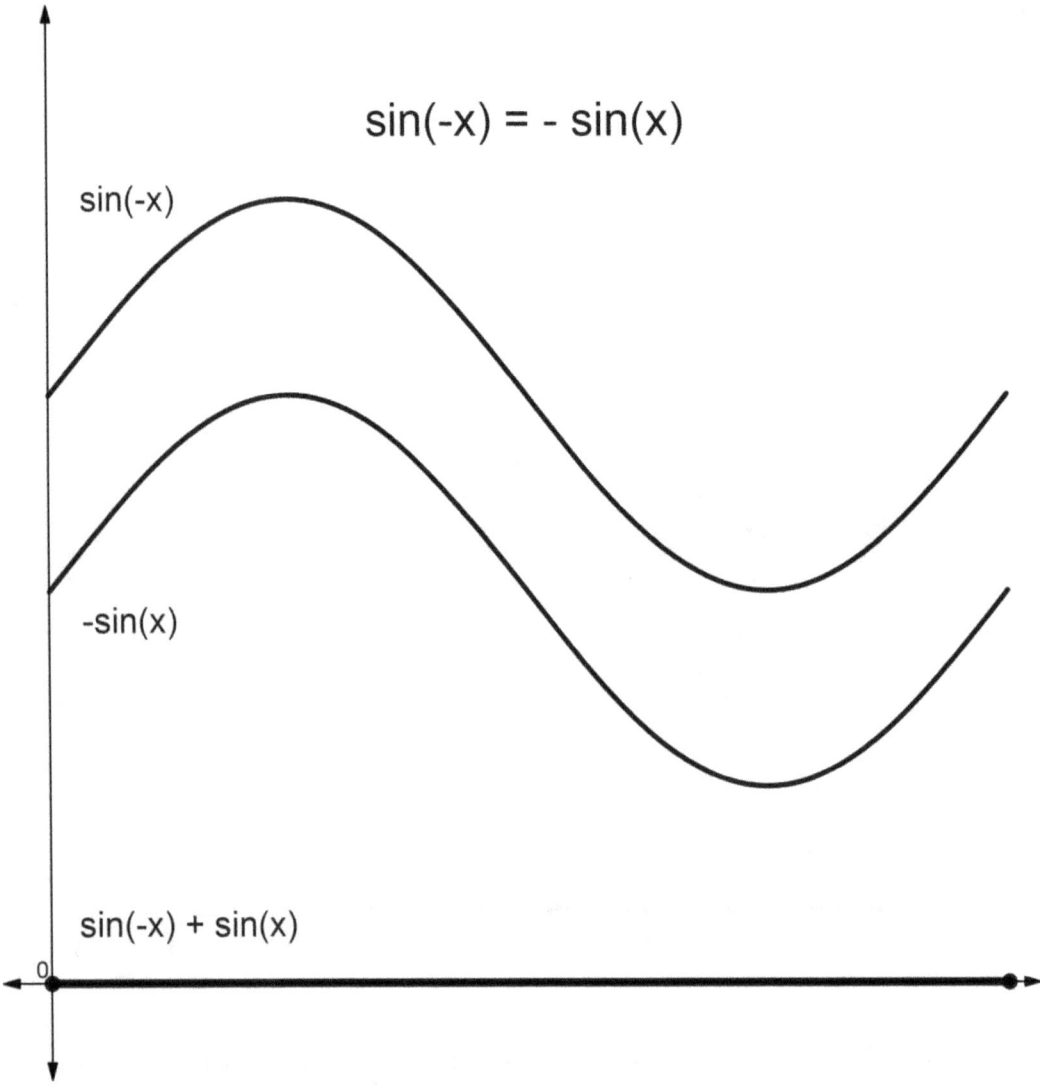

Figure 27 shows the graphs of sin(-x) and -sin(x) over the range 0 to 2π. They repeat after that. As before, the graphs have the same shape. The horizontal axis contains the subtraction of the two graphs, sin(-x) + sin(x). (We recall that subtracting a negative quantity turns it into the addition of a positive quantity.) Since the vertical component of a line on the horizontal axis is "0", sin(-x) must equal -sin(x). **Sin(-θ) = -sin(θ)** is another common identity used in trigonometric calculations.

Now that the Cosine and Sine identities are out of the way, we can prove that "sinh(iθ)", or ".5[e^(iθ) - e^(-iθ)]", equals "i sin(θ)". The equation for the imaginary hyperbolic Sine wave is repeated below:

54. $\sinh(i\theta) = .5[e^{(i\theta)} - e^{(-i\theta)}]$

The "$e^{(i\theta)}$" and "$e^{(-i\theta)}$" terms of Equation 54 are Euler Formula imaginary unit circles and the "θs" are radian values from where an imaginary unit circle crosses the horizontal x-axis to a point on the perimeter of the circle. We recall the convention for measuring radians on a real or imaginary circle is that the value is positive if measured in a counterclockwise direction and negative if measured in a clockwise direction. So the only difference between the two imaginary unit circles in Equation 54 is that the first one, $e^{(i\theta)}$, is measured in a counterclockwise direction and the second one, $e^{(-i\theta)}$, is measured in a clockwise direction.

Let's substitute for the terms in Equation 54. From Euler's Formula:

55. $e^{(i\theta)} = \cos(\theta) + i \sin(\theta)$ and

56. $e^{(-i\theta)} = \cos(-\theta) + i \sin(-\theta)$

We know from the Cosine and Sine identities that:

57. $\cos(-\theta) = \cos(\theta)$ and

58. $\sin(-\theta) = -\sin(\theta)$

Substituting Equations 57 and 58 into Equation 56:

56. $e^{(-i\theta)} = \cos(\theta) - i \sin(\theta)$

Subtracting Equation 56 from Equation 55:

59. $e^{(i\theta)} - e^{(-i\theta)} = \cos(\theta) + i \sin(\theta) - \cos(\theta) + i \sin(\theta)$

Simplifying the right-hand side of Equation 59:

60. $e^{(i\theta)} - e^{(-i\theta)} = i \sin(\theta) + i \sin(\theta) = i\, 2\sin(\theta)$

Multiplying both sides of Equation 60 by .5:

61. $.5[e^{(i\theta)} - e^{(-i\theta)}] = i \sin(\theta)$

From Equation 54:

62. $\sinh(i\theta) = .5[\, e^{(i\theta)} - e^{(-i\theta)}\,] = i \sin(\theta)$

Or, dropping the Euler Formula notation:

63. $\sinh(i\theta) = i \sin(\theta)$

We have found that mathematically an equation exists that allows imaginary Sine waves to be created in the future. We have derived Equation 63 by utilizing what we know about Euler Formula imaginary unit circles. The reader can see that the imaginary Sine wave came through the calculation intact. But moving from Euclidian (Euler Formula) spacetime to hyperbolic spacetime, where we exist, caused the problematic real cosine wave of the Euler Formula to vanish.

We discussed earlier that radians, θ, can represent either arc length or time. If we measure it as the distance of a point on the unit circle from the origin, or where the unit circle intersects the horizontal x-axis (e.g. - Figure 20), it is an arc length. If we alternatively measure it as the time it takes to get to a point on a unit circle, it is time. Accordingly, we let the θ in Equation 63 represent time:

64. $\sinh(iT) = i \sin(T)$

where the "θ" of Equation 63 is time, T.

Equation 64 tells us that imaginary Sine waves, $i \sin(T)$, exist in imaginary hyperbolic time, $\sinh(iT)$. The left-hand side of Equation 64 represents imaginary hyperbolic time, or the hyperbolic future because iT represents imaginary time, or the future, and because iT is contained in a function describing hyperbolic spacetime. The right-hand side of Equation 64 represents an imaginary Sine wave, or a brainwave. If only there were a way to manipulate $\sinh(iT)$, imaginary hyperbolic time. If we could do that, we could create any brainwave we wanted to in the hyperbolic future.

We can do this. Remember from Equation 51 that V = iT, which says that we can visualize the future after we shut off our sensors. Let's insert Equation 51 into the left-hand side of Equation 64:

65. $\sinh(V) = i \sin(T)$

Equation 65 tells us that, once we shut off our sensors and move to the hyperbolic future, we can create any event we desire, by visualizing it. We have shown in Equation 51 that we can visualize the hyperbolic future after we have shut down our sensors. And, once we have done so, Equation 65 tells us we can create brainwaves by visualizing the events that create them.

Visualization, V, in the left-hand side of Equation 65 represents the visualization of events in the hyperbolic future. But the event brainwaves can be expressed in either distance or time because they are Sine waves (or Cosine waves) measured in radians, as we have seen. Let's therefore change time, T, on the right-hand side of Equation 65 back to the general case, radians, we have been using above:

66. $\sinh(V) = i \sin(\theta)$

We have seen above that, when dealing with waves, "θ" can be expressed as:

67. $\theta = 2\pi f t + \alpha$

where:
"f" is the frequency of the wave
"t" is the duration of the wave
"α" is the phase of the wave.

Substituting Equation 67 into the right-hand side of Equation 66:

68. $\sinh(V) = i \sin(2\pi f t + \alpha)$

Equation 68 tells us that, when we shut off our sensors and move to the hyperbolic future, sinh(V), we can visualize, or create, imaginary brainwaves, i sin(2πf t + α). Since OBEs are inside events not involving our sensors, we also experience in the hyperbolic future the brainwaves we have created there. There is no need to let the future brainwaves trickle down the time escalator to the present where they will be picked up by our sensors.

When we visualize events in the hyperbolic future, do we visualize the individual components on the right-hand side of Equation 68? In other words, we are in the hyperbolic future trying to create an event. Do we visualize the individual frequencies, f, durations, t, and phases, α, the event creates? I suppose this is possible, but I don't believe this is what actually happens.

I believe brainwave creation in the hyperbolic future is probably automatic as we visualize events. In other words, visualizing an event in the future using the left-hand side of Equation 68 likely *means* we are creating the brainwave parameters (f, t, α) contained on the right-hand side of Equation 68 whether or not we are aware of it. Asking for, or visualizing, an event to be created may mean that we automatically manipulate the parameters of the brainwaves that comprise the event. While in the future, or imaginary time, part of hyperbolic spacetime as seen in Equation 68, we probably create any event simply by asking for it, or visualizing it in our mind's eye. This automatically creates the event's frequencies (f), durations (t) and phases (α). During my OBEs, I have never consciously specified the frequencies, durations and phases of the event I creating. I simply ask for it to occur by imagining or visualizing it and then it occurs.

When we are in the hyperbolic future, I believe the part of us that is there (our consciousness) **is** a wave. Equation 68 supports this. Equation 68 says that once we enter the hyperbolic future by shutting off our sensors, we **are** imaginary Sine waves. As we then visualize moving around in the hyperbolic future by varying V, or iT, our brainwave frequencies, f, durations, t, and phases, α, are automatically created by our movement and the brainwaves we create are the "footprints" of our movements.

When I am having an OBE, I move like I am, and everything else is, a wave. It is very common for me to want to leave the room (usually bedroom) I am in and see what is going on outside the room. How do I leave the room? Do I open the door? No. I simply will myself to go through the room's door or even its wall and it happens. I am a wave and the wall or door is a wave. Those that have had a physics class know that solid objects can't pass through each other but waves can. There is nothing in the future that is solid. When the future trickles down to our sensors, it becomes solid in the present. I believe that our sensors *cause* the present to be solid.

We recall that the radius, "r", of the Euler Formula imaginary circle was the amplitude of the wave created by moving around the imaginary circle. The amplitude of Equation 68's Sine wave is understood to be "1" because every variable or function in mathematics has an implied "1" in front of it if there is no other number there, so:

69. $1*\sinh(V) = 1 * i \sin(2\pi f t + \alpha)$

There is nothing in Equations 68 or 69 that allows for an imaginary amplitude greater than "1". What gives? The amplitude of a wave is proportional to its energy. The larger the amplitude, or height, of a wave, the greater the wave's energy. Anyone who has stood in the ocean has experienced this. If a wave with a small height hits you, you can just stand there as the wave washes around you. But if you stay in the water long enough, sooner or later a wave with a bigger height might arise which knocks you over. Think of the surfers in Hawaii who surf 20-foot waves.

Below, we are going to discuss the possibility, and I believe the very real probability, that we can leave the present and go into the future and create outside events (as opposed to inside event OBEs) that may combine with and be experienced with outside events others have created in the future. I believe not being able to create imaginary waves with amplitudes greater than "1" may be nature's way of not letting us have too much power when we create future events that will also be experienced by others. If a person could go into the future and create an outside event with a very large amplitude (energy), that event could conceivably "move mountains" when it trickled down to the present. Instead, nature may say "one person, one vote".

Not being able to control the amplitude of an imaginary brainwave does not prevent us from creating imaginary brainwaves as described by Equation 68 above. The amplitude of a wave only affects the height (energy) of the wave, not its frequency, duration, phase or anything else. It is like listening to music. The amplitude of the music waves affects only the volume. If the music you are listening to is too loud and someone asks you to turn it down (teenagers, anyone?), the only thing that changes is the sound level, not the words, frequency, duration, pitch, etc. Only the volume.

An alternative explanation is that the amplitude of the Sine wave for an event we create in the future may be a function of the amount of time we are able to visualize the event being created. We will see below that two (or more) identical Sine waves with amplitudes of "1" can combine and reinforce each other. After they combine, the frequency, duration, and phase of the resultant wave will be the same as the original waves. But one thing will be different: the amplitude of the combined wave will be the sum of the amplitudes of its component waves (e.g.: 1 + 1 = 2).

So if we are able to continue to focus on a particular event we want to create while we are in the hyperbolic future, that will be the same as sending out a stream of identical Sine waves which will then combine to have an amplitude that is the sum of the amplitudes of the individual waves you have created. In other words, the amplitude of a Sine wave you create in the future may be a function of the time you spend focusing on its creation. The longer you are able to focus, the greater the amplitude of the wave.

CHAPTER 11

HAVING AN OBE

I believe the best time to have an OBE is at night. At night, your sensors are already turned down or off. The sun goes down and your eyes receive less light. There is much less traffic noise to activate your hearing sensor. And so on. This is why OBEs are likely to occur at night while you are sleeping. It is easier to tunnel out of your sensors in the first place. And it is less likely that your sensors will be activated and pull you out of your OBE. This is especially the case if you are able to sleep in a relatively quiet, stimulus free environment. Under these conditions, with practice, it may be possible to focus and "train" your sensors to stay turned off when a fairly small change in your environment takes place.

Modern technology may also be able to play a role in inducing and/or maintaining OBEs. For example, noise-cancelling earphones are able to block out noise in your environment. I was given a pair of noise-cancelling earphones as a gift. I was amazed at the almost absolute silence when I put them on. For example, one day I was sitting at my desk in my office. My desk was just a few feet away from a window and the window was not far from my neighbor's yard. All of a sudden, my neighbor appeared just outside with a chainsaw. He began cutting limbs from a tree with the chainsaw blaring. He was probably only thirty or so feet away from me. With the chainsaw blaring, I could not hear a thing and I certainly could not focus on the work I was trying to do. I put on my noise-cancelling earphones and the sound of the chainsaw was completely eliminated! We will describe how noise-cancelling earphones work below.

What other sensors besides our ears can we turn off in order to help us experience an OBE? Many people use sleep masks to block all input to their eyes while they are sleeping. Your taste sensor is turned off or down automatically when you are not eating. Placing yourself in an environment that is free of strong smells or odors will turn down your smell sensor.

Your touch sensor (skin) is a little more problematic because that sensor warns us of threats to our bodies (e.g. - heat, cold, insects) and therefore may be harder to turn off. However, there is a device called a sensory deprivation tank, isolation tank or floatation tank that turns down or off your skin and other sensors, including your vision and hearing sensors. A sensory deprivation tank is a tank filled with Epsom salt and water heated to the outer skin temperature. A person experiences minimal sensory stimulation when inside the tank. The tank contains enough water and Epsom salt for the person entering it to float on their back.

The downside to using a sensory deprivation tank is that they are expensive, costing from around $10,000 to more than $30,000. An alternative to buying a sensory deprivation tank is to rent one. There are places (think Tanning Salons) that will rent time in a flotation tank for around $50 to $100 dollars per hour. Some of these places have membership programs.

The sensory deprivation tank was actually invented by a person named John Lilly. He began to develop the tank in the 1950s as a dark, soundproof enclosure half-filled with saline water meant to deprive a person of all external stimuli (sensor input) in order to free the conscious mind from the body (have an OBE). His 1972 book, "The Center of the Cyclone" describes his experimentation with sensory deprivation tanks. (It is interesting that a cyclone can be described as a funnel.) In addition, the 1980 movie hit, "Altered States", was based on his sensory deprivation tank experiences.

In order to further disconnect his body from the outside world (i.e. - shut off his sensors) while he was in a sensory deprivation tank, Mr. Lilly took an anesthetic called "ketamine". He was already familiar with the drug because he had taken it to relieve his chronic migraine headaches. He spent hours inside the sensory deprivation tank, carefully noting his reaction to various doses of the drug.

A purpose of an anesthetic, including ketamine, is to shut off the body's sensors during surgery. Ketamine got its start in Belgium in the 1960s as an anesthesia medicine for animals. The FDA approved it as an anesthetic for people in 1970. It was used in treating injured soldiers on the battlefields in the Vietnam War. Unlike other anesthetics, ketamine doesn't slow breathing or heart rate, so patients don't need to be on a ventilator to receive it. Ketamine is currently being studied as an antidepressant for persons who aren't able to obtain relief from other medications.

I have not used ketamine or any other dissociative drug to help induce an OBE and I don't recommend doing so to anyone else. Although taking drugs *may* cause the user to have a dissociative experience, they also can, and probably will, distort the very reality the user is trying to experience (think LSD trip). Furthermore, these drugs are illegal and can be addictive.

I wrote a computer program that puts a funnel on the monitor. The funnel can be animated to make it appear that you are moving down it. The speed of the animation can be changed by the user of the program. I have found that, when I view the animation for fifteen or twenty minutes a day for two or three days, an OBE is likely to occur at night in the next few days. What happens is that I begin to see the computer-generated funnel in my mind while I am sleeping. My subconscious mind then says to me, "Hey, here's a funnel. You can go down it." I then go down the funnel and have an OBE.

I think there is a lot more that can be done. For example, it is possible to measure a person's brainwaves in real time with a compact, relatively inexpensive electroencephalograph which produces an "electroencephalogram" ("EEG"). The machine's output could be fed into a computer program such as the program which creates the animated funnel described above. Suppose you are viewing the animated funnel while your brainwaves are being simultaneously recorded. Suppose also that the frequency of your brainwaves is at a relatively high level and you do not have an OBE after "going down" the funnel for a fair amount of time. The computer which is generating the animated funnel and simultaneously receiving the EEG input could then lower the frequency, or speed, at which the user is moving through the tunnel. This should cause the frequency of the user's brainwaves to decline towards a level at which an OBE may occur. The process of a person's brainwave frequency being caused to match the frequency of what is being stared at on a computer monitor is called "entrainment".

The computer/EEG setup we describe represents a feedback system. If a person's measured brainwave frequency is higher than that during which an OBE occurs, the program lowers the frequency of the funnel. On the other hand, if the person's measured brainwave frequency is lower than that during which an OBE occurs, the program raises the frequency of the funnel. In each case, entrainment causes the person's measured brainwave frequency from "going down" the computer-generated funnel to move towards the frequency at which an OBE is likely to occur.

No one knows the brainwave frequency or the range of brainwave frequencies at which OBEs occur. This is an area that is ripe for research. But by using a feedback system as described above, it may be possible to "zero in" on a frequency at which an OBE occurs.

Your sensors are designed to keep you safe. They warn you of impending danger. For example, your eyes and ears sense oncoming traffic and tell you it is not safe to cross the street. Your sense of smell warns you of toxic chemicals in the environment. And your sense of touch senses heat which keeps you away from fire. Sooner or later, one of your sensors will "go off" while you are having an OBE and your OBE will end.

CHAPTER 12

OUTSIDE EVENTS AND CREATING THE FUTURE

We have been discussing OBEs, which are inside events. Because inside events do not involve input from our sensors, the brainwaves involved in inside events are both created and experienced in the future by the person having the OBE. And because our sensors are turned off during an inside event, they do not allow us to interact with the brainwaves created by other persons. Inside events happen *inside* you. That is why they are called "inside" events.

Outside events come to us through our sensors. Accordingly, outside events are events that can involve multiple people. When we create an outside event, we create brainwaves that describe the entire event, including the brainwaves that will be experienced by other people who participate in the event. Outside events involve things that take place *outside* of us. That is why they are called "outside" events. Suppose, for example, we want to create an outside event that involves playing in a baseball game. You create an outside event in which your team wins the game. This outside event will clearly affect the other players in the game, especially the players on the other team who are also trying to win the game.

Creating both inside and outside events involves turning off our sensors and moving to the future. The difference between inside and outside events involves what happens when we *get to* the future. When we create an inside event, or OBE, it affects only us. Only we can experience the OBE. No sensors required. On the other hand, once we get to the future, we can also create an outside event that could affect many other people. Sensors required! How do we create an outside event?

Let's create an outside event where we win the baseball game discussed above. We turn off our sensors and move to the future. (We now know that the "future" is the "hyperbolic future.") Using the horizontal time escalator, we create and plant brainwaves three hours into the future on the escalator. We create these future brainwaves in the same way we created the OBE brainwaves discussed above. But these brainwaves contain the brainwaves that will be experienced by all of the players and spectators when our team wins the game because our visualization, V, affects the outcome of the game that they will all experience. This brainwave "package" then moves towards the present: three hours away, two hours away, one hour away, and so on. Eventually, the time escalator brings the brainwaves to the present, they are picked up by the players' and spectators' sensors, and our team has won the game!

The reader may be thinking that the game might take four hours to play. How can we plant brainwaves three hours away creating the outcome of the game? Good question. The answer is that, as the expected end time of the game changes, the person who created the brainwaves can turn off his or her sensors, move back into the future and adjust the brainwave package's position on the time escalator.

What a deal! Our team will never lose another baseball game! World Series, here we come! It is intuitively (and historically) understood that this can't happen and never has happened. What have we missed? What we have not considered is that everyone, all players from each team and all spectators, have access to the time escalator. All of these persons can turn off their sensors and visualize the game's outcome by simply *thinking* about the outcome of the game. They can create and deposit their own brainwaves on the time escalator which reflect the outcome *they* want to occur. What a mess! Brainwave stacked upon brainwave. Watch where you step!

The brainwaves will work it out amongst "themselves" on the time escalator. We recall that brainwaves are Sine waves. From physics, we know that Sine waves can combine with one another. We discussed this above when we considered the amplitude of our brainwaves. And we mentioned above that the branch of mathematics known as "Fourier Analysis" allows any mathematical curve describing any event to be broken down into a series of Sine waves. Some unidentical Sine waves on the escalator will cancel each other out. Some will add together. Some will partially add together and partially cancel each other out. The result will be a game outcome that *nobody is certain of* even though they placed their own desired outcomes on the time escalator. As the game is played, the brainwaves combine and the expected outcome is more certain. But not until the game is over do the deposited brainwaves reveal the true outcome because the players and spectators can deposit new brainwaves as the game is played. Ninth inning rallies happen all the time.

How do brainwaves combine? Figure 28 shows brainwaves that cancel each other out:

Figure 28

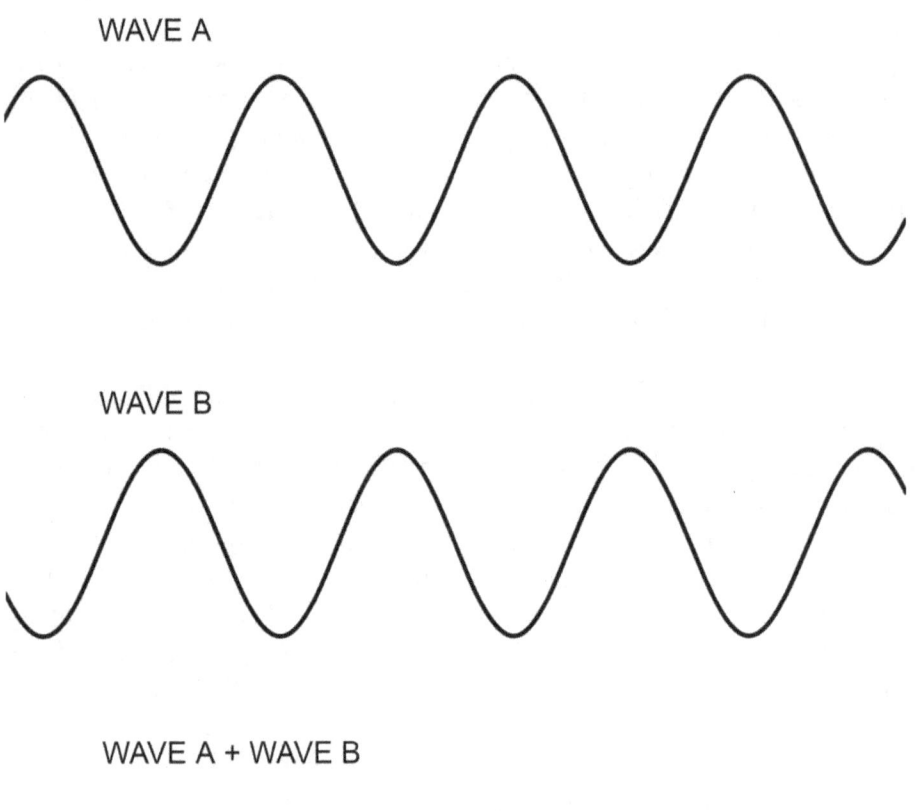

Continuing our baseball game analogy, the top wave in Figure 28 is from a player on team A that is up to bat and who visualizes hitting a home run. The bottom wave in Figure 28 is from the pitcher on team B that visualizes striking out the player who wants to hit the home run. Each of these waves represents part of the events that the hitter and pitcher are trying to create in imaginary time, or the future. The events are mutually exclusive. The hitter cannot simultaneously hit a home run and strike out.

Notice that every time the top wave, the batter's wave, reaches a peak, or highest value, the bottom wave, the pitcher's wave, reaches its trough, or lowest value. As we have seen, the high and low points of a wave are determined by the wave's amplitude. The amplitude is a measure of the strength of the wave. We are assuming that the amplitude of the hitter's and pitcher's waves is the same and that the trough amplitudes of each wave are the negative of the peak amplitudes. In other words, let's say that the peak amplitudes of each wave are 1 unit. The trough amplitudes of each wave are therefore -1 unit. This is the description of a normal wave.

We add two waves by summing their amplitudes. We can see from Figure 28 that whenever the amplitude of the hitter's wave is at its peak or 1 unit, the amplitude of the pitcher's wave is at its trough, or -1 unit, and vice versa. The sum of the amplitudes of each wave is zero at every point as seen by the straight line in Figure 28. The straight line means that the waves have disappeared. The amplitude of the straight line is '0' and it therefore has no frequency. The hitter's and pitcher's waves have solved the problem of being mutually exclusive by cancelling each other out!

Put another way, the hitter's and pitcher's waves cancel each other out because they are the exact "opposite" of one another. When they cancel out, their waves vanish. The hitter's and pitcher's waves don't "know" each other. They don't know anything about the events they represent. Cancelling out happens automatically. Once the cancelling occurs, the events represented by the cancelled waves disappear.

As we have learned, the "phase difference" between two waves is a measure of how much one is offset from the other in either time or distance. If the waves are measured in time units, the phase difference is measured in time units. If the waves are measured in arc lengths, the phase difference is measured in arc lengths. In Figure 28, the hitter's and pitcher's waves are said to be "180 degrees out of phase". This just means what we have already seen: the waves are the complete opposite of one another and have completely cancelled each other out. They are gone.

This is an example of what is called "destructive interference". (The word "interference" just means that the waves interact. In the case under discussion, the interaction is "destructive".) It is exactly what occurs when sound waves are cancelled by the popular noise-cancelling earphones that we discussed above. The headphones intercept an incoming sound wave from your environment. Sophisticated electronics contained in the headphones then create the opposite, 180-degree out of phase, sound wave which combines with the incoming wave, cancelling each wave out. Noise-cancelling earphones could just as well be called "sound-cancelling earphones" because that is just what they do.

The mutual exclusivity of the hitter's home run wave and the pitcher's strike out wave have been cancelled out. This means the hitter is not going to hit a home run *and* he is not going to strike out. What is actually going to happen will be determined by other waves that the hitter, pitcher, players and spectators have deposited on the time escalator. It is important to note that the two waves cancel each other out because the only difference between them is that they are 180 degrees out of phase. If that were not the case, the sum of the two waves would be non-zero and something else would have happened during the at-bat. Walk? Fly-out? Single?

Figure 29 shows two waves that demonstrate "constructive interference":

Figure 29

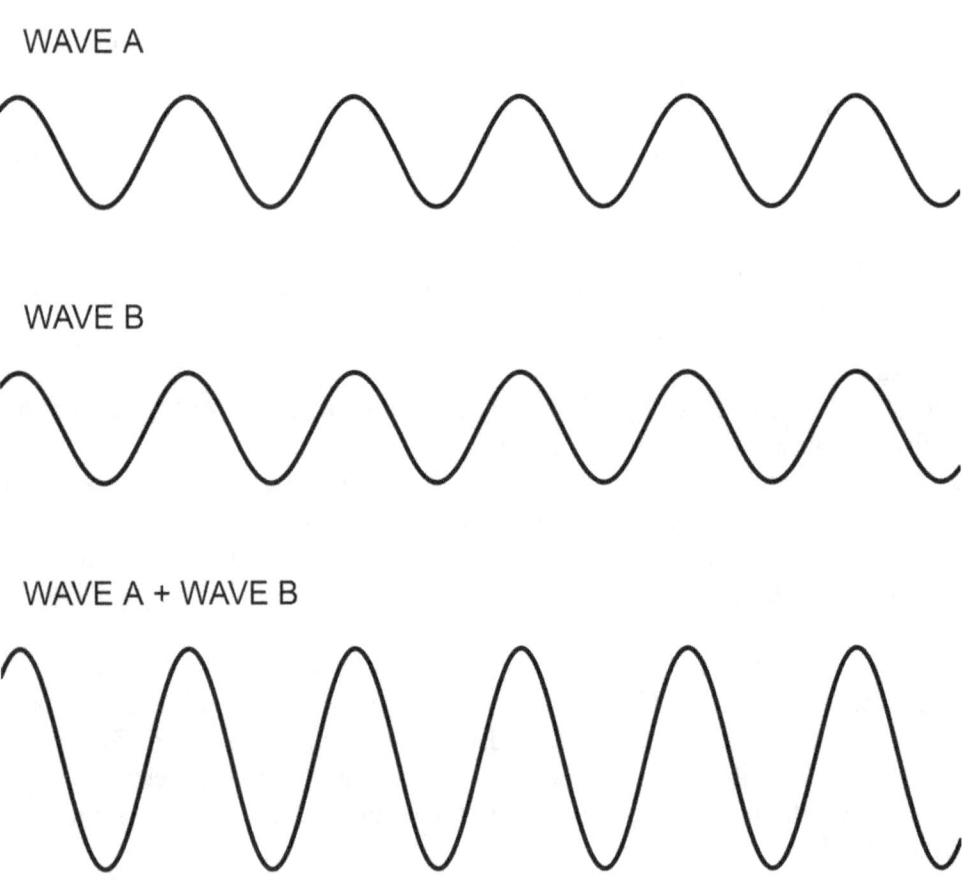

In Figure 29, the top wave, Wave A, is from a player who visualizes that the umpires will be unbiased. The bottom wave is from another player (on either team) who also visualizes that the umpires will be unbiased. Notice that wave A and Wave B are identical with A's peaks corresponding to B's peaks and A's troughs corresponding to B's troughs. This means that the two waves are "in phase" or have a "0-degree" phase difference.

The bottom wave in Figure 29 results when we add the two player's waves together. Notice that the amplitude of the combined wave is greater than the amplitude of either player's wave. In fact, the amplitude of the combined wave is twice the amplitude of either player's wave because the amplitude of wave A is equal to the amplitude of wave B. The addition of same-frequency waves that are in phase with one another is an example of "constructive interference". As can be seen from Figure 29, constructive interference creates a stronger wave from its component waves.

Constructive interference strengthens the likelihood of unbiased umpires because the two players have created identical "unbiased" waves in the future and deposited them on the time escalator. Since the amplitude of a wave represents its energy, the combined "unbiased" wave is twice as strong as either separate "unbiased" wave. This means that the two mutually inclusive "unbiased" waves have reinforced each other.

We have seen that 180-degree phase shifts result in wave cancellation or destructive interference. And we have seen that if the phase shift is 0 degrees, the amplitudes are added together through constructive interference making the underlying wave for the particular frequency even stronger. We have focused thus far for simplicity on 0-degree phase shifts and 180-degree phase shifts. But there are an infinite number of possible phase shifts, or wave offsets, that can occur between those two extremes (e.g. – 50-degree phase shifts, 90-degree phase shifts, etc.). In fact, it is likely that most waves injected into imaginary time by persons visualizing future events will be offset from one another in either distance or time by some amount. That is one reason the same frequency wave does not necessarily mean the same thing to different persons. The same frequency waves of each person may be offset from one another, conveying different meanings to them.

Figure 30 shows two same-frequency waves that are offset, or phase shifted, by 145 degrees:

Figure 30

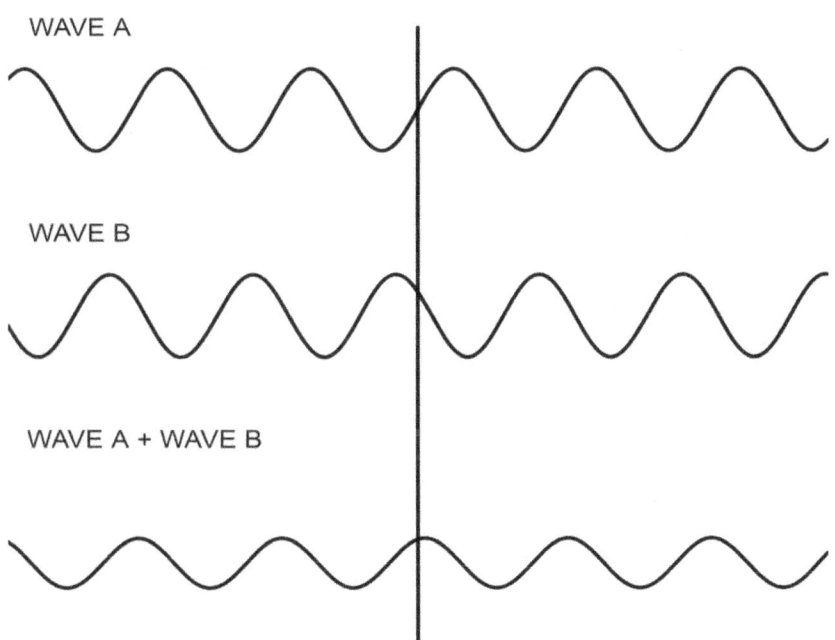

The vertical line in Figure 30 shows that when Wave A is approaching a peak, Wave B has rounded a peak and is on its way to a trough. Combining Waves A and B creates a wave, Wave A + Wave B, that is approximately at a peak. The 145-degree phase shifted Wave A + Wave B has not been cancelled out as in a 180-degree phase shifted wave (Figure 28). Neither does the amplitude of the combined wave equal twice the amplitude of the individual waves (Figure 29). The 145-degree phase shifted Wave A + Wave B has an amplitude that is less than the amplitude of either Wave A or Wave B (whose amplitudes are the same). We can see from this that Wave A + Wave B is "in between" the 180-degree phase shifted wave seen in Figure 28 and the 0-degree phase shifted wave seen in Figure 29.

Our demonstration thus far has concerned waves that have the same frequencies for each person participating in the event (e.g. - pitcher and hitter). Waves that have different frequencies also interact. In fact, it is likely that **all** of person A's frequencies will interact with **all** of person B's frequencies in resolving any conflict between events being activated during the same future time unit. We have focused thus far on common frequencies because they are mathematically straightforward. But there are an infinite number of possible frequencies that can combine constructively or destructively.

Figure 31 illustrates the combination of two waves, Wave A and Wave B, that have different frequencies:

Figure 31

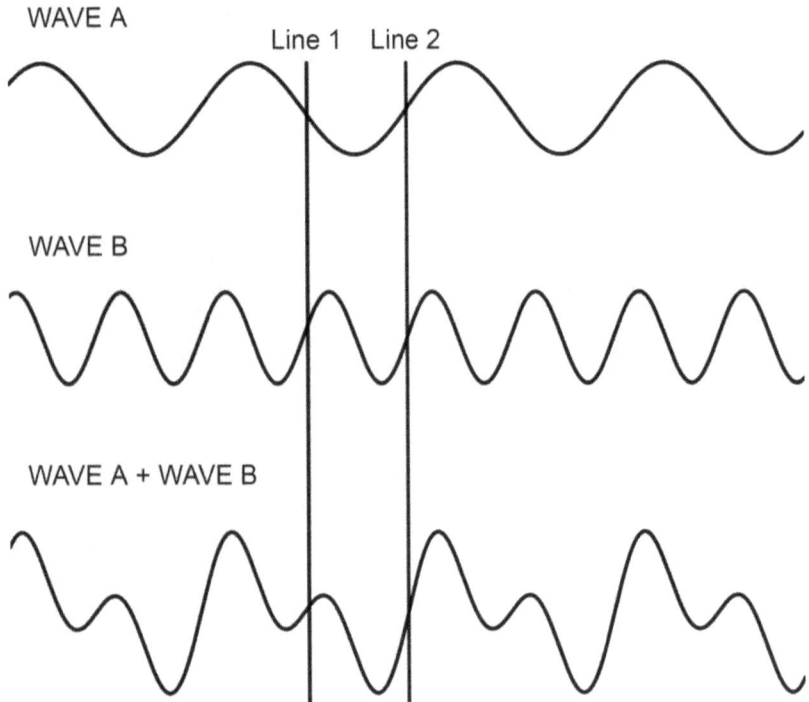

Figure 31 shows two brainwaves, Wave A and Wave B, that have different frequencies. Wave B's frequency is twice the frequency of Wave A. The waves combine to form Wave A + Wave B. Notice that Wave A + Wave B has a major peak followed by a minor peak followed by a single trough before the cycle repeats. There is no phase difference between Waves A and B. They are "in phase". Sometimes Waves A and B or combine constructively and sometimes they combine destructively. Note that the difference from major peak to trough of Wave A + Wave B, a measure of wave amplitude, is greater than that for either Wave A or Wave B.

Vertical Line 1 shows destructive interference between Wave A and Wave B. Wave A is declining from a peak and Wave B is moving towards a peak. The net result is that Wave A + Wave B is approaching a minor peak. Vertical Line 2 illustrates constructive interference between Wave A and Wave B. Both Wave A and Wave B are approaching peaks and their combination, Wave A + Wave B is approaching a major peak.

Figure 32 illustrates the combination of two brainwaves, Wave A and Wave B, that have different frequencies as well as a phase shift between them:

Figure 32

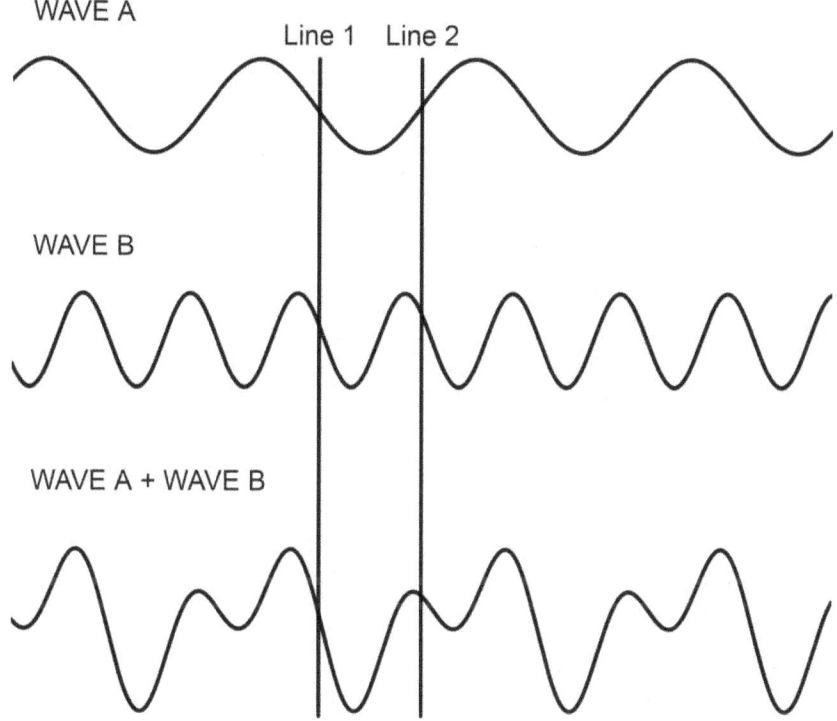

In Figure 32, Wave B's frequency is twice Wave A's frequency as in Figure 31. This time, however, the waves are phase shifted with Wave B being 145 degrees "ahead" of Wave A. Because of the phase shift, Line 1 now shows constructive interference while Line 2 now shows destructive interference. Also because of the phase shift, the minor peak occurs before the major peak instead of after it as shown in Figure 31. Vertical Line 1 shows constructive interference because both waves have passed a peak and are declining. Vertical Line 2 shows destructive interference because Wave A is moving towards a peak while Wave B has rounded a peak and is moving towards a trough. As in Figure 31, the difference from major peak to trough of Wave A + Wave B, a measure of wave amplitude, is greater than that for either Wave A or Wave B.

Thus far, all of our examples have involved phase shifts and/or frequency differences between Waves A and B. Waves A and B have always shared the same amplitude. Remember from our previous discussion that amplitude, or the "height", of a wave is a measure of its energy, or strength. In another ocean wave example, imagine being in a small boat on the ocean. You would not be troubled if the height of the ocean waves were 6 inches. But imagine how you would react if the waves' height suddenly changed from 6 inches to 6 feet!

Any wave can be described by its amplitude, frequency and phase (as well as its duration). Figure 33 illustrates two waves, Wave A and Wave B, that have different amplitudes, frequencies and phases:

Figure 33

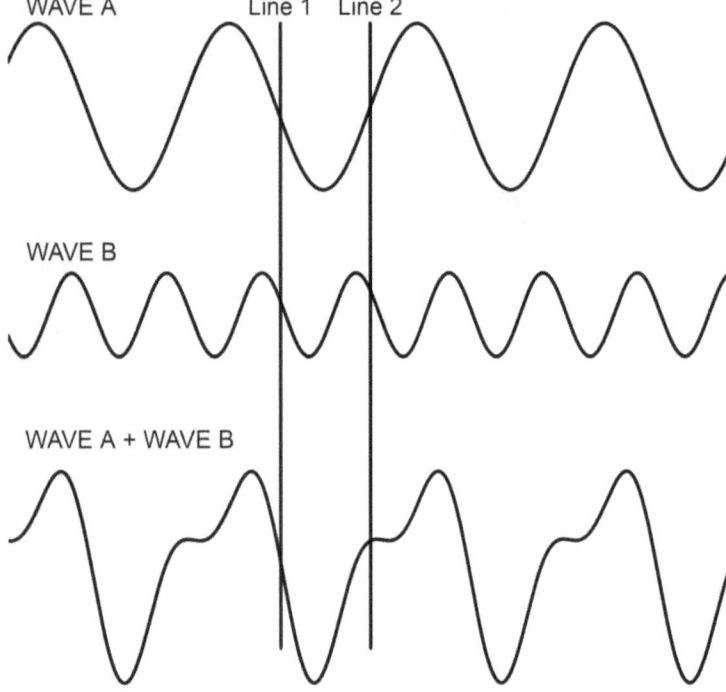

In Figure 33, Wave A's amplitude is twice the amplitude of Wave B while Wave B's frequency is twice the frequency of Wave A. Wave B's phase is 145 degrees ahead of Wave A. As before, the combination of Wave A and Wave B, or Wave A + Wave B, enjoys both constructive and destructive interference. For example, vertical Line 1 shows constructive interference because both Wave A and Wave B are declining from a peak towards a trough. On the other hand, vertical Line 2 shows destructive interference because, while Wave A is increasing from a trough to a peak, Wave B is decreasing from a peak to a trough.

Notice that the double peaks of Figures 31 and 32 have largely disappeared. Figure 33 has more of a plateau than a peak. This is because the higher amplitude of Wave A has "overpowered" Wave B and prevented Wave B from creating a discernable minor peak in Wave A + Wave B. Notice also that the difference from trough to peak of Wave A + Wave B, a measure of its amplitude, is larger than that of either Wave A or Wave B. Finally, Wave A + Wave B's frequency is the same as Wave A's frequency.

We can see from the above that the combination of only two persons' brainwaves of different amplitude, frequency, phase and duration can, and usually does, result in a "child" wave that is significantly different than its "parent" waves. All waves, regardless of amplitude, frequency and phase, will combine through constructive interference, destructive interference or a combination of both to create a new wave that shares the "DNA" of, but is not the same as, the original waves. The wave created will depend on the amplitude, frequency and phase attributes of the original waves. It is not hard to see from the above examples that there are an unlimited number of different child waves that can result from the combination of only two parent waves.

Things can get very complicated when you consider the large number of people who may be attempting to visualize or activate events at the same future time and place. Each event someone attempts to activate at a future time and place will be combined with all events attempting to be activated at the same future time and in the same future place by others. What actually takes place at the selected future time and place will be the result of the combination of all participants' brainwaves.

A group of people acting in concert as if they have one mind is a real phenomenon and is called "Team Flow". On October 4, 2021, a team of researchers from Toyohashi University of Technology, Tohoku University and the California Institute of Technology reported in the journal "Neuroscience" that they had found evidence of brainwave team flow. According to the researchers, "Team flow is experienced when team players get "in the zone" to accomplish a task together. Great teams experience this psychological phenomenon, from sports to music bands and even professional work teams. When teamwork reaches the team flow level, one can observe the team perform in harmony, breaking their performance limits." **The researchers found a unique signature of team flow: increased beta and gamma brainwaves. The researchers determined that, during Team Flow, the participants experienced synchronized brainwave activity or the simultaneous combination of identical brainwaves.**

Changing sports from baseball to football, consider the example of two football teams about to play. Each team has 30 players. Each of the players on Team A and 29 of the players on team B visualizes his team winning the game. The frequency of this visualization is 20 hz for all players on each team who have visualized the outcome (30 players on Team A and 29 players on Team B). If the amplitude of each player's 20 Hz wave is 1, then, through constructive interference after all of Team A members' visualizations are added up, the amplitude of the "Team A wave" will be 30, or 30*1. However, because only 29 of Team B's members have visualized victory, the amplitude of the "Team B wave" will only be 29 through constructive interference. Because both teams winning is mutually exclusive, their team waves will be offset and undergo destructive interference. The resultant wave will be Team A's wave with an amplitude of 1 (30 team A - 29 team B), and Team A will win the game. The reader has probably witnessed a sporting event in which the underdog team has become "of one mind" and upset the favorite. I know I have.

It is possible for a person to not try to activate an event at a given time and place in the future. In this case, the event experienced by the person "going along for the ride" will be the event "chosen" for him as the mutual exclusivity and other characteristics of future events at the given time and place chosen by others is worked out.

In summary, we have postulated that all brainwaves of any frequency, amplitude, phase or duration injected into any future imaginary time unit and place will be combined with one another in some manner. The mutually exclusive waves will cancel each other out and disappear from any future event that filters down through time to the present on the time escalator. The mutually inclusive waves will reinforce each other to strengthen the impact of those waves on events that reach the present. The events that result from the combination of the participants' brainwaves will almost certainly not be the same as the original future events the individual participants attempted to activate. The process of combining brainwaves is a dynamic one as the time escalator marches from the future to the present.

If outside events are created in the future by us, it follows that people must have the ability to *regularly* go to the future to "plant" them on the time escalator. We have seen herein that about one-in-ten persons has reported having an OBE and that there are not a lot of people (yet) who have reported that they regularly experience them. It follows that there must be other ways in which we can create future outside events that will march through time along the time escalator until they arrive at the present to be captured by our sensors and experienced by us in the present. Creating outside events in the future, which often, but not always, involve other people, requires us to turn our sensors off in order to visualize the event. But once created, we rely on our sensors so we can experience the outside events in the present after they have arrived on the time escalator from the future.

Our brains contain what is called the "Default Mode Network", or "DMN". The DMN is the part of our brain that is responsible for abstract and self-directed thought, including **visualization and imagining the future**. When the DMN is "off", or inactive, we observe the world created by our sensors. When it is "on", or active, we are able to visualize the future with our imaginations. Try visualizing some future event you want to experience. You can do that fairly easily. The very act of beginning to visualize the event will cause your DMN to turn your sensors off, allowing the visualization to take place. How do you know that your sensors are turned off? The answer is simple. If you can imagine the future event, your sensors have been turned off. We are either in the present or in the future. Visualizing a future event causes our DMN to turn off our sensors, if only for a brief period of time.

Recall our example of viewing an Olympic athlete (gymnast, downhill skier, etc.) on television just before they compete in an event. They often make a sort of blank stare into space as they visualize themselves being successful in their event. What they are doing is turning on their DMN, which allows them to visualize and place their desired outcome of the event (winning it) on the time escalator. The very act of *thinking* about a future event turns on the DMN long enough for a person to place brainwaves for that event on the time escalator. We don't need to keep our sensors turned off for OBE lengths of time to move to the future and create outside events.

We have seen that turning off our sensors is a matter of degree. If you want to focus on the road while driving your car, you can "tune out" minor distractions caused by your children in the back seat. If you want to imagine some future event, you can turn off your sensors via your DMN long enough for the visualization to occur. And if you want to have an OBE, you can, with practice and the right environment, turn off your sensors for the longer period of time required to have one.

Is there anything else that leads us to the future? I believe so. What happens when we fall asleep? Our sensors turn off. Our sensors turning off is why we fall asleep. We mentioned above that we are unaware of clock time when we are sleeping. That is because our sensors are inactive and therefore unable to bring clock time to us. If our sensors turning off causes us to fall asleep, then "where" are we when we dream? We are in the future. When we dream, our eyes are moving. This is called "REM" sleep. "REM" stands for "Rapid Eye Movement". We are essentially watching the event that we are visualizing in the hyperbolic future when we dream.

After I had started having OBEs, there was a time when I applied for a job. I felt the job would be a good fit for me and I really wanted it. I had interviewed twice with the person who would be my manager if I were to be hired. This person was responsible for making the hiring decision. I was told by her that there were multiple candidates for the position. I had not heard back about the job for a few days. I worried that I would not be offered the position. One night I had a dream that was not associated with an OBE. I dreamed that I was speaking to the hiring manager and I told her I really wanted the job.

The next morning, I got a call from the hiring manager who offered me the position and asked if I could start the next day! I accepted the position and told her I would see her the next day. When I showed up the following morning, the first thing I did was meet with the hiring manager in her office. As we were talking, she told me: "You know, I was going to hire another candidate for this position but two nights ago I had a dream that you should be the one I hired." The very night I dreamed I spoke to the hiring manager and expressed my enthusiasm for the job is the same night she dreamed that she should change her mind and hire me instead of another candidate! Is this evidence that my dream activated a future event in which the hiring manager would pick me for the position? To me it is.

Daydreaming and dreams differ in at least two respects. First of all, the person having a daydream consciously picks the subject to daydream about while the dreamer, at least consciously, does not. Secondly, as we have seen, daydreams are relatively short compared to dreams. It seems that the shortness of daydreams compared to dreams may be offset by their targeted content.

We have postulated that we experience outside events in the present along the horizontal events-axis as they arrive there from the future. As we experience present events, they become past events. We can make modifications and corrections to past events by moving to the future where we can "tweak" the results of past events which will then trickle down the time escalator until they arrive at the horizontal events-axis as present events. We can't change past events, but we can create new events which adjust for them.

We have also postulated two ways in which someone can move to the future in order to create, or activate, a future event. One way is through visualization (including daydreaming and thinking), which can occur in short intervals courtesy of our DMN. The other way is through dreams, which are spontaneous in nature.

This would be a good place to consider what causes me to dream about future outside events that will come true after I have had an OBE?" All outside events start out as future events when they are visualized in the future. And all outside events count down in time on the time escalator until they become present events on the horizontal events-axis. All events, both inside (OBEs) and outside events, we experience start out in the future. I believe that, while I am having an inside event OBE in the future, I may "run across" an outside event on the time escalator as it is trickling down to the present. The event registers in my consciousness and comes back to me as a dream after my OBE ends. Since I dream about the outside events right after my OBE has ended and since the outside events are still traveling to the present on the time escalator when I have my dreams, the events have not yet occurred when I dream about them. It is like driving down an interstate highway and you pass a bus full of people. After you have passed the bus, you stop at a gas station five miles down the road. As you are filling your tank, the bus passes by. You knew this was probably going to happen because you passed the bus in the past, five miles up the road.

Outside Events Summary

Our bodies are a collection of sensors that receive signals from "outside" of us and then pass them along to our consciousnesses as brainwaves. When the brainwaves impact our consciousnesses, we recognize them as outside events. Every outside event is a package of brainwaves that have come to our consciousnesses by way of our sensors. Different events will have a different combination of brainwaves. If the brainwaves produced by one event are different from those produced by another event, we recognize those events as different.

We create an outside event by visualizing it. The visualization causes brainwaves describing the event to be deposited on the time escalator. Outside events usually involve multiple people. Accordingly, multiple people who will witness the event may deposit brainwaves in that event as it trickles down the time escalator. Different participants in an event will usually want different event outcomes, resulting in many competing brainwaves. The result is that the brainwaves deposited on the time escalator will work the differences out. Brainwaves of different amplitudes, phases, frequencies and durations will be combined. Mutually exclusive brainwaves will be cancelled out. Mutually inclusive brainwaves will be added together. When the outside event reaches the present, its brainwaves will be picked up by the participants' sensors and they will experience the event. The events that result from the combination of the participants' brainwaves will almost certainly not be the same as the original future events that the individual participants attempted to activate.

In order to create an outside event in the future, we must go there. We have postulated two ways a person can get to the future to create an outside event: visualization and dreaming. Any person who even thinks about a future outside event he or she will participate in will unconsciously enter the future and deposit brainwaves into the event on the time escalator.

CHAPTER 13

REQUIREMENTS FOR CREATING OUTSIDE EVENTS

What are the requirements for creating outside events in the future and what are the chances we can meet those requirements? First of all, we have to be able to go to the future. We know that the future is imaginary. It is not solid like the present is. We can only imagine it in our minds. It is a virtual certainty that we can go to the future because we experience OBEs. I have done it many times. And as we have mentioned, approximately one-in-ten persons have experienced an OBE. But OBEs are inside events, not outside events.

The primary way we go to the future to create an outside event is through visualization, which includes as little as thinking about an event. Equation 51 above, $V = iT$, shows us that, in the hyperbolic spacetime that we exist in, our visualizations, V, take us to imaginary time, iT, or the future. In addition, as discussed above, we have postulated that there is another way to visit the future besides visualizing it: dreams. Certainly, anyone who has had a dream come true after having the dream had to have been in the future while having the dream (coincidences aside).

Secondly, we have to be able to create events when we get to the future. The experience of having an inside event OBE involves visualizing events in the future which we experience in the future. And there are many examples of persons visualizing future events while daydreaming in order to cause their wishes to become true in the present. We talked about athletes such as Olympic gymnasts who appear to stare into blank space just before competing. And we mentioned that during "Team Flow", athletic teams and other groups of people, including office workers, experience synchronized brainwaves. Why does this happen if not to create a desired outcome in the future which will come true in the present?

Equation 68 above, $\sinh(V) = i \sin(2\pi f t + \alpha)$, tells us that, when we shut off our sensors and move to imaginary hyperbolic time, "$\sinh(V)$", we can create imaginary brainwaves, "$i \sin(2\pi f t + \alpha)$" of any frequency, f, duration, t, and phase, α. In a nutshell, Equation 51, $V = iT$, gets us to imaginary time, the future, by visualizing it as our sensors are turned off and Equation 68, $\sinh(V) = i \sin(2\pi f t + \alpha)$, allows us to create brainwaves by visualizing events after we have arrived there. We have further postulated that visualizing an event while we are in the hyperbolic future automatically creates the brainwave frequencies, durations and phases describing the event. When we move to the hyperbolic future, we *become* the Sine waves described in Equation 68.

Thirdly, we have to deposit event brainwaves on the time escalator. The time escalator is a metaphor for the processes that bring brainwaves created through visualization in the future to us in the present. We know that time flows from the future to the present. Any physicist will tell you that time does not flow backward from the present to the past. It only goes one way: from the future towards the present. We also know that we become Sine waves when we turn off our sensors and move to the hyperbolic future. Our movements through the hyperbolic future literally create Sine waves. When we create brainwaves (Sine waves) in the future, they have to go somewhere. Where? They march from the future to the present along what we have been calling the "time escalator".

Fourthly, the brainwaves of event participants have to combine in the future to create one reality when they reach the present. We spent some time above discussing how Sine waves can combine. The amplitudes, frequencies, durations and phases of multiple Sine waves can mathematically come together. Waves that "clash" will cancel each other out through destructive interference while waves that reinforce each other will add together through constructive interference. We mentioned above that any mathematical curve representing any event is a collection of Sine waves. And what are Sine waves? They are brainwaves.

And Fifthly, after the brainwaves "roll off" the time escalator into the present, they have to impact our sensors so that we can experience the event they represent. We know that brainwaves come to us through our sensors. This is an unchallenged, established fact. Our sensors are the vacuums that "sweep up" outside events and present them to our consciousnesses as they roll off the time escalator into the present.

Given the above, I would say there is a pretty good chance that we create at least some outside events by visualizing them in the hyperbolic future.

CHAPTER 14

EVIDENCE FOR OUTSIDE EVENTS

I believe the theory of how, and where, OBE's occur discussed herein is sound. As noted above, it is estimated that 1 in 10 persons have experienced OBE's. I have experienced many of them. The OBE theory is built on how I have an OBE and what happens during and after (dreams) I have an OBE. On the other hand, the theory proposed on activating future events in imaginary time is more speculative than the OBE theory. But I thoroughly believe this is what happens. As we discussed above, there is scientific evidence for "Team Flow", which is what happens when teams, often athletic teams, synchronize their brainwaves to accomplish a common goal such as winning an event. Synchronizing brainwaves is an example of the constructive interference of a team's brainwaves which causes the amplitude (power) of the common brainwaves to be able to overcome the brainwaves of the team's competitor(s), resulting in the team's success at obtaining their common goal.

I would like to cite two personal examples of creating an event in the future that moved down the time escalator to become true in the present. The first event is one I cite to the reader in all humility. I am a septuagenarian. I have never been a great athlete but I have been exercising and lifting weights since I was around ten years old. You can go into just about any gym and see a bunch of strong young guys lifting weights. There are older guys in the gym that are not as strong. They may have been as strong as the younger guys but not surprisingly time has taken its toll. Not so with me. I am as physically strong now as I was forty years ago. In fact, I am stronger and weighing 185 pounds can machine bench press over 400 pounds.

How can this possibly be? Well, I know that events can be "planted" on the time escalator in the future. And I know the time escalator brings the planted events to us in the present. So I visualize lifting a relatively heavy weight in the future. When the time escalator brings the visualization to the present, the event happens and I lift the weight. Put another way, we know that the future consists of waves. There is nothing concrete or heavy in the future. Everything is a wave there. Waves don't have mass. The future only becomes solid, or heavy, when the time escalator brings it to the present. So I visualize lifting the weight in the future while it is a light, easy-to-manipulate, wave. I do this when it is just a second or two away from the present. When the time escalator brings the brainwaves to the present, my visualization becomes reality.

The other example I would like to cite is as follows. One day, I was working outside at my house. It was hot. I was tired. I was trying to finish the job so I could go back inside and cool off. All of a sudden, a good friend sent me a text message. There was nothing wrong. He just wanted to chat. I responded to his message. A minute later, he sent me another message. I did not want to be unkind and tell him I was too busy to message with him, so I replied to his message. Then he texted me again and I replied again. This was repeated several more times and I finally **thought to myself**, "Leave me alone!" A few minutes later, my friend texted me yet again. He asked me, "What does LMA mean?" I told him I had no idea. He then said, "Then why did you text 'LMA' to me? Does it mean 'Leave Me Alone'?" I told him I had sent him no such text. To confirm this, I checked my phone.

There was the text to him: "LMA". I was shocked! Somehow my *thought* for him to "Leave me alone!" ended up on my and his phones in a text message saying "LMA". Apparently, just thinking this unkind thought projected the event onto the time escalator to be subsequently seen by him (and me) when it trickled down to the present as a text message! I only found out later that "LMA" is actual slang for "Leave me alone!" So even if I would have consciously texted him that I did not want to converse, there was no way I would have known to text "LMA". Furthermore, that is not the way I treat people. (We're still friends.)

CHAPTER 15

FINAL THOUGHTS

The author wishes to thank those who have read this book. We are on a journey together in our voyage through space, time, the present and the future. Keep wondering. Keep searching. Keep asking "Why?". And keep burrowing.

Any reader with questions or comments is welcome to contact the author at:

treilorbanks@gmail.com

AUTHOR BIOGRAPHY

The author is a graduate of Stanford University and the Massachusetts Institute of Technology (M.I.T.). He is a member of Mensa. He has been interested in physical fitness his entire life and is classified as an elite weightlifter. He is married and has two adult children. He and his wife Linda live in Spring, Texas. His academic interests are mathematics, physics, electronics, graphics programming and neuroscience. He enjoys teaching mathematics to young people.